EXPLORATIONS IN BEGINNING AND INTERMEDIATE ALGEBRA USING THE TI-82/83/83 PLUS/85/86 GRAPHING CALCULATOR

THIRD EDITION

Deborah Jolly Cochener
Austin Peay State University

Bonnie MacLean Hodge
Austin Peay State University

THOMSON

BROOKS/COLE

Australia • Canada • Mexico • Singapore • Spain • United Kingdom • United States

Printed in The United States

3 4 5 6 7 8 9   09 08 07

Printer: Globus Printing

ISBN-13: 978-0-534-40644-8

ISBN-10: 0-534-40644-0

For more information about our products, contact us at:
Thomson Learning Academic Resource Center
1-800-423-0563

For permission to use material from this text,
contact us by:
Phone: 1-800-730-2214
Fax: 1-800-487-8488
Web: http://www.thomsonrights.com

Asia
Thomson Learning
5 Shenton Way #01-01
UIC Building
Singapore 068808

Australia/ New Zealand
Thomson Learning
102 Dodds Street
South Street
Southbank, Victoria 3006
Australia

Canada
Nelson
1120 Birchmount Road
Toronto, Ontario M1K 5G4
Canada

Europe/Middle East/South Africa
Thomson Learning
High Holborn House
50/51 Bedford Row
London WC1R 4LR
United Kingdom

Latin America
Thomson Learning
Seneca, 53
Colonia Polanco
11560 Mexico D.F.
Mexico

Spain/ Portugal
Paraninfo
Calle/Magallanes, 25
28015 Madrid, Spain

Tell me, and I'll forget.
Show me, and I may not remember.
Involve me, and I'll understand.

-Native American Saying

✤ ✤ ✤ ✤

TABLE OF CONTENTS

Basic Calculator Operations

Graphically Solving Equations and Inequalities

Graphing and Applications of Equations in Two Variables

Stat Plots

INTRODUCTION OF KEYS

Unit Title	TI-82 Keys	TI-83/83plus Keys	TI-85/86 Keys
#1: Getting Acquainted With Your Calculator, p.1	ON/OFF 2nd ▲ ▼ (to darken/lighten) MODE ENTER CLEAR (-) abs ( ) parentheses √ x^2 ∧ QUIT INS DEL π ENTRY ANS	ON/OFF 2nd ▲ ▼ (to darken/lighten) MODE ENTER CLEAR (-) CATALOG MATH ►NUM 1: abs(( ) parentheses √ x^2 ∧ QUIT INS DEL π ENTRY ANS	ON/OFF 2nd ▲ ▼ (to darken/lighten) MODE ENTER CLEAR (-) CATALOG CUSTOM abs ( ) parentheses √ x^2 ∧ QUIT/EXIT INS DEL π ENTRY ANS
#2: Fractions and the Order of Operations, p. 11	MATH 1:►Frac	MATH 1:►Frac	CUSTOM►Frac
#3: The STOre and Table Feature, p.19	STO► X,T,θ ALPHA : Y = TABLE	STO► X,T,θ,n ALPHA : Y = TABLE	STO► x-VAR ALPHA : y(x) = TABLE (TI-86) TI-85 (no table) use EVAL or evalF
#4: Scientific Notation, p. 29	MODE (SCI)	MODE (SCI)	MODE (SCI)
#5: Rational Exponents and Radicals, p. 35	MATH 4:$\sqrt[3]{}$ 5:$\sqrt[x]{}$	MATH 4:$\sqrt[3]{}$ (5:$\sqrt[x]{}$	CUSTOM $\sqrt[x]{}$
#6: TI-83/83plus/85/86 Complex Numbers, p. 39	DOES NOT APPLY	i MATH ► CPX MODE Real► a + bi	(a,b) = a +bi CPLX

Unit Title	TI-82 Keys	TI-83/83plus Keys	TI-85/86 Keys

#7: Graphical Solutions: Linear Equations, p.45	Y =
	 WINDOW
	 TRACE
	 GRAPH
	 CALC
	     5:intersect
	 QUIT
	 ZOOM
	     6:ZStandard
		Y =
		 WINDOW
		 TRACE
		 GRAPH
		 CALC
		     5:intersect
		 QUIT
		 ZOOM
		     6:ZStandard
			GRAPH
			 y(x) =
			 RANGE
			 TRACE
			 GRAPH
			 MATH/ISECT
			 EXIT
			 MORE
			 GRAPH/ZOOM
			     ZSTD

#8: Graphical Solutions: Absolute Value Eq., p.55	No New Keys	No New Keys	No New Keys

#9: Graphical Solutions: Quadratic and Higher Degree Equations, p.59	ZOOM
	     1:ZBox
	     2 :Zoom In
	 CALC 2:root
		ZOOM
		     1:ZBox
		     2 :Zoom In
		 CALC 2:zero
			GRAPH/ZOOM
			     BOX
			     ZIN
			 GRAPH/MATH
			 ROOT

#10: Applications of Quadratic Equations, p.69	CALC
	     4:maximum
	     1:value
	     (EVAL X)
	 ZOOM
	     8:ZInteger
		CALC
		     4:maximum
		     1:value
		     (EVAL X)
		 ZOOM
		     8:ZInteger
			GRAPH/MATH
			 FMAX
			 GRAPH/MORE/
			 EVAL
			 ZOOM: ZINT

#11: Graphical Solutions: Radical Equations, p.75	No new keys	No new keys	No new keys

#12: Graphical Solutions: Linear Inequalities, p.81	Y-vars
	     1:Function..
	     1:Y1
	 TEST menu
		VARS
		     Y-VARS
		     1:Function..
		     1:Y1
		 TEST menu
		 Graph Style Icon
			y1: USE ALPHA
			     KEY TO
			 ENTER
			
			 TEST menu
			 Graph Style Icon

#13: Graphical Solutions: Absolute Val. Ineq., p.87	No new keys	No new keys	No new keys
#14: Graphical Solutions: Quadratic Inequalities, p.91	WINDOW ▶FORMAT AxesOn/AxesOff	FORMAT AxesOn/AxesOff	GRAPH FORMAT AxesOff
#15: Graphical Solutions: Rational Inequalities, p. 97	MODE Connected/Dot	MODE Connected/Dot	GRAPH FORMAT DrawDot

#16: How Does the Calculator Actually Graph, p.103	GRAPH
	 ZOOM
	     6:ZStandard
	 WINDOW
		GRAPH
		 ZOOM
		     6:ZStandard
		 WINDOW
			GRAPH
			 GRAPH/ZOOM
			     ZSTD
			 RANGE/WIND

Unit Title	TI-82 Keys	TI-83/83plus Keys	TI-85/86 Keys
#17: Preparing to Graph: Calculator Viewing Windows, p.107	ZOOM 4:ZDecimal 5:ZSquare 8:ZInteger	ZOOM 4:ZDecimal 5:ZSquare 8:ZInteger Graphing Icon	GRAPH/ZOOM ZDECM ZSQR ZINT Graph Icon - 86 GRAPH/SELCT
#18: Where Did the Graph Go?, p.119	No new keys	No new keys	No new keys
#19: Functions, p.123	DRAW 1:ClrDraw 3:Horizontal 4:Vertical 8:DrawInv VARS	DRAW 1:ClrDraw 3:Horizontal 4:Vertical 8:DrawInv VARS	GRAPH/MORE DRAW CLDRW VERT DrInv
#20: Discovering Parabolas, p.131	No new keys	No new keys	No new keys
#21: Translating & Stretching Graphs, p.137	No new keys	No new keys	No new keys
#22: Exponential and Logarithmic Functions, p.143	No new keys	No new keys	No new keys
#23: Predict-a-Graph, p.151	No new keys	No new keys	No new keys

Unit Title	TI-82 Keys	TI-83/83plus Keys	TI-85/86 Keys

#24: Statistics: Plotting Paired Data, p.155	STAT
	  1:Edit
	  4:ClrList
	STAT PLOT
	  1:Plot 1
	MEM
	   2:Delete
	     3:List
	ZOOM
	  9:ZoomStat
		STAT
		  1:Edit
		  4:ClrList
		STAT PLOT
		  1:Plot 1
		MEM
		
		4:ClrAllLists
		
		ZOOM
		
		9:ZoomStat
			STAT
			  EDIT
			    CLRxy
			
			  DRAW
			    SCAT
			    xyLINE
			
			  ZOOM
			  ZDATA

#25: Frequency Distributions, p.165	STAT/CALC
	  1:1-Var Stats
	  3:SetUp
		STAT/CALC
		  1:1-Var Stats
			STAT/CALC
			  OneVa(TI-86)
			STAT/DRAW
			  HIST

#26: Line of Best Fit, p.175	STAT/ CALC
	  3:SetUp
	
	5:LinReg(ax + b)
	VARS
	
	5:Statistics/EQ
	  7:RegEQ
		STAT/CALC
		  2:2-Var Stats
		4:LinReg(ax + b)
		VARS
		
		5:Statistics/EQ
		  1:RegEQ
			STAT/CALC
			  TwoVa(TI-86)
			  LINR
			VARS
			  STAT
			    RegEq

PREFACE

This unique workbook/text provides the student the opportunity for guided exploration of elementary and intermediate algebra topics using Texas Instruments graphing calculators. Keystroking guides are provided for the TI-82, TI-83, TI-83plus, TI-85, and TI-86 calculators. This enables the instructor to use the text in a classroom that requires any of the above listed calculators, or in the classroom where calculators are mixed. The text is intended for use as a supplemental text to a CORE classroom text, and is therefore arranged by topics. This enables the instructor to assign the appropriate *Explorations Unit(s)* that correlate(s) with the topic under discussion in the classroom. **The text is not meant to be worked in sequential order. Each unit has one or more prerequisite units. *Only* the prerequisite units are required for student success in working the assigned unit.** This allows the use of this ancillary text with *any* core course textbook. Charts that correlate the concepts from textbook sections with specific *Explorations* units are available in the *Instructor Resources* section of the text specific web site for Brooks Cole (www.mathematics.brookscole.com).

Changes in the Third Edition

▶ The TI-83/83plus are the base calculators used in the text, with changes noted for the TI-82, TI-85, and TI-86 graphers.

▶ The key correlation charts have all been moved to the front of the text, following the Table of Contents, to provide a quick reference for students.

▶ The Table of Contents lists the prerequisite unit(s) for each unit and specifies the correlating concept. This allows both the instructor and the student to better link the appropriate calculator features to algebraic concepts.

▶ More writing exercises have been interspersed throughout the text and are clearly marked by the ✍ icon. These exercises are more clearly defined in this edition and are often separated as individual problems to encourage students to respond in more detail.

▶ This edition contains fewer units than previous editions. An effort has been made to combine related concepts in a single unit rather than interspersing a concept through several different units.

▶ Units have been reorganized to better reflect current pedagogy. For example, the TABLE feature is introduced in Unit 3 in conjunction with the use of the STOre feature, allowing the early integration of the use of tables.

Features

*Each unit provides guided exploration of a topic. Units are not meant to be done in numerical order, but rather according to concept. A correlation chart within the Table of Contents correlates units to traditional text topics. Prerequisite units are listed in the Table of Contents as well as at the beginning of each unit.

*The text requires NO instruction - but serves as a workbook for the student. Answers to questions/exercises within a unit appear at the end of the unit.

*Some units may be used in place of classroom instruction or to introduce a concept. This is designated within the Table of Contents.

*A key correlation chart that shows the units in which keys are introduced is provided immediately following the Table of Contents.

*The workbook may be used over a period of more than one semester/quarter as the student progresses through his/her mathematics sequence.

*As an institution changes textbooks, this supplement need not be changed.

*A Troubleshooting section is provided. It contains common student errors as well as explanations of the error screens students most often encounter.

*Units are written in a manner that enables the student with little or no algebraic experience to read and explore independently.

*The units provide springboards for both classroom discussion and further investigation either individually or as a class.

Integration of technology into the mathematics classroom has grown significantly with the introduction of the graphing calculator. These calculators have brought relatively inexpensive technology into the students' hands. However, instructors are now faced with the problem of integrating the technology without sacrificing course content. These activities will enable the students to develop algorithms typically found in elementary and intermediate algebra courses, improving both their understanding and retention of material.

The workbook was written with the belief that students who are active contributors in the classroom increase their own understanding and their long term retention of material. Of primary importance, however, is the fact that the units have provided the means and the opportunity for students to create their own mathematics.

Organization

The workbook is divided into four sections - Basic Calculator Operations, Graphically Solving Equations and Inequalities, Graphing and Applications of Equations in Two Variables, and Stat Plots.

UNITS #1-6: These units introduce the student to the calculator and its capabilities in performing computational tasks.

UNITS #7-15: These units provide the link between the algebraic processes traditionally taught in elementary or intermediate algebra courses and the graph associated with the algebraic equations and inequalities. These units can be worked before the text formally introduces graphing. However, instructors may wish to assign Units 16-18 prior to Unit 7.

UNITS #16-23: These units introduce the student to graphing as a formal process. It is in these units that adjusting view screens (WINDOWS) and interpreting graphs is emphasized. Emphasis in this section is on functions.

UNITS #24-26: The statistical units are *not* meant to provide a comprehensive coverage of the subject; they *are* intended to give the student the keystroking information necessary to introduce them to the basic capabilities of their calculator.

PREFACE TO THE STUDENT

This workbook was developed with YOU in mind. It is written in a style that is easily understood by students at varying levels of mathematical proficiency. The text provides the keystroking information necessary to use the calculator as a tool in your mathematics class. **The units are not meant to be worked in order**. The Table of Contents provides the prerequisites for each unit. Those prerequisites are also specified at the beginning of the unit. Those are the *only* units required prior to completion of a unit assigned by your instructor. The following are suggestions for using the workbook:

STUDY SKILLS

***READ** and follow the instructions *slowly* and *carefully*. Pay close attention to detail.

*Do not skip the questions marked with a writing icon (✎). These questions help you bring your thoughts together and put the mathematics into language that is meaningful to YOU.

*Keep a log of those calculator techniques that have proven helpful.

*In this log, also write any questions and/or comments that occur to you as you work through the units. Form a study group with your classmates to discuss these entries. **WHEN YOU DISCUSS MATHEMATICS, YOU LEARN MATHEMATICS.**

*Highlight keystroking information/directions that are helpful to you.

*The calculator is a **TOOL** for learning and doing mathematics. If a result does not seem reasonable, always double check your reasoning, keystrokes and screen display.

*Answers appear at the end of each unit.

*The units are correlated to algebraic concepts - be sure you work the prerequisite units.

*Keep this workbook; it will serve as a <u>personalized</u> reference for future courses.

We hope that by learning to use the graphing calculator your confidence in your abilities to create and DO mathematics increases. Experiment frequently with the keys and menu options that are not covered in the workbook. Last, but not least, HAVE FUN!

ACKNOWLEDGEMENTS

We are grateful and appreciative for the input of time and expertise by those who reviewed this workbook:

  Jennifer Dollar, Grand Rapids Community College
  Laurette Foster, Prairie View A & M

We appreciate the guidance and support of Jennifer Huber, Senior Acquisitions Editor, Rachael Sturgeon, Senior Assistant Editor, Leah Thomson, Senior Marketing Manager, and Carrie Dodson, Editorial Assistant at Brooks/Cole publishing. We would also like to thank our husbands, David and Blaine, and our children, Sherah, Blaire, and Trey, for their support and love. They have always believed in us and it is they who made the dream become reality. Our families' contributions to the editing, reviewing, wording, and the testing of material has been invaluable.

UNIT 1
GETTING ACQUAINTED WITH YOUR CALCULATOR

*This unit is a prerequisite for all other units in the text. Answers appear at the end of the unit.

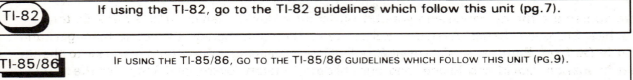

TI-82   If using the TI-82, go to the TI-82 guidelines which follow this unit (pg.7).

TI-85/86   IF USING THE TI-85/86, GO TO THE TI-85/86 GUIDELINES WHICH FOLLOW THIS UNIT (PG.9).

Touring the TI-83/83plus

Take a few minutes to study the TI-83/83plus graphing calculator. The keys are color-coded and positioned in a way that is user friendly. Notice there are dark blue, black, and gray keys, along with a single yellow key and a single green key.

Dark blue keys: On the right and across the top of the calculator are the dark blue keys. At the top right are four directional cursor keys. These may be used to move the cursor on the screen in the direction of the arrow printed on the key. The four arithmetic operation symbols are also in dark blue. Notice the key marked **ENTER**. This is used to activate entered commands, thus there is no key on the face of the calculator with the equal sign printed on it. Below the screen are five keys labeled **[Y=]**, **[WINDOW]**, **[ZOOM]**, **[TRACE]**, and **[GRAPH]**. These keys are positioned together below the screen because they are used for graphing functions. The TI-83plus has an additional blue key labeled **APPS**. It is used to access the finance menu and the CBL/CBR capabilities of the calculator.

Black keys: The majority of the keys on the calculator are black. Notice the **[X,T,θ,n]** key in the second row and second column. It will be used frequently in algebra to enter the variable **X**. The **ON** key is the black key located in the bottom left position.

Gray keys: The twelve gray keys that are clustered at the bottom center are used to enter digits, a decimal point, or a negative sign.

Yellow key: The yellow key is the **2nd** key located in the upper-left position. To access a symbol in yellow (printed above any of the keys) first press the yellow **[2nd]** key, and then the key BELOW the symbol (function) to be accessed. For example, to turn the calculator **OFF** notice that the word **OFF** is printed in yellow above the **ON** key. Therefore, press the keys **[2nd] [ON]** to turn the calculator off. These keystrokes are done *sequentially* - not simultaneously. Throughout this book the following symbolism will be used: symbols that appear on the key will be denoted in brackets, [ ], whereas symbols written above the key will be denoted in < >. Thus, the previous command for turning the calculator off would appear as **[2nd] <OFF>**. The symbols [ ] or < > will cue you <u>where</u> to look for a command - either printed on a key or above it.

Green key: Alphabet letters (printed in green above some of the keys), or any other symbol and/or word printed in green above a key are accessed by first pressing the green **ALPHA** key and then the key below the desired letter/symbol/word. The keystrokes are sequential.

Catalogue feature: Press [2nd] <CATALOG> to display an alphabetical list of available calculator operations. Use the [▲] and [▼] cursor keys to scroll through this list. Operations may be accessed by placing the pointer adjacent to the operation and pressing [ENTER].

Note: The TI-83/83plus has an Automatic Power Down (APD) feature which turns the calculator off when no keys have been pressed for several minutes. When this happens, press [ON] to access the last screen used.

Let's Get Started!

Turn the calculator on by pressing [ON]. If the display is not clear, press [2nd] [▲] to darken the screen, or [2nd] [▼] to lighten the screen. Notice that when the [2nd] key is pressed, an arrow pointing up appears on the blinking cursor.

To ensure the calculator is in the desired mode, press [MODE]. All of the options on the far left should be highlighted. If not, use the [▼] to place the blinking cursor on the appropriate entry and press [ENTER]. Exit MODE by pressing [CLEAR]. This accesses the home screen which is where expressions are entered. Press [CLEAR] until the screen is cleared except for the blinking cursor in the top left corner.

Integer Operations

When entering integers on the calculator, differentiation must be made between a subtraction sign and a negative sign. Notice that the subtraction sign appears on the right side of the calculator, with the other arithmetic operations. The negative sign appears to the left of the [ENTER] key and is labeled (-).

Example: Simplify: -8 - 2

Keystrokes:
[(-)] [8] [-] [2] [ENTER]

Remember: the symbols enclosed in brackets appear on a calculator key.

Screen display:

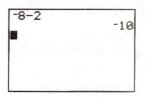

Observe the difference in size and position of the negative sign as compared to the subtraction sign.

TI-85/86   TI-85/86 USERS TURN TO "ABSOLUTE VALUE" IN THE GUIDELINES (PG. 10).

Absolute Value

Absolute value is accessed by pressing [MATH] [▶](NUM) [1:abs( ] or through the [CATALOG]. The absolute value operation is displayed as **abs(** . When absolute value is accessed, a left parenthesis appears with the abbreviation *abs*. You must close the parenthesis to ensure correct evaluation.

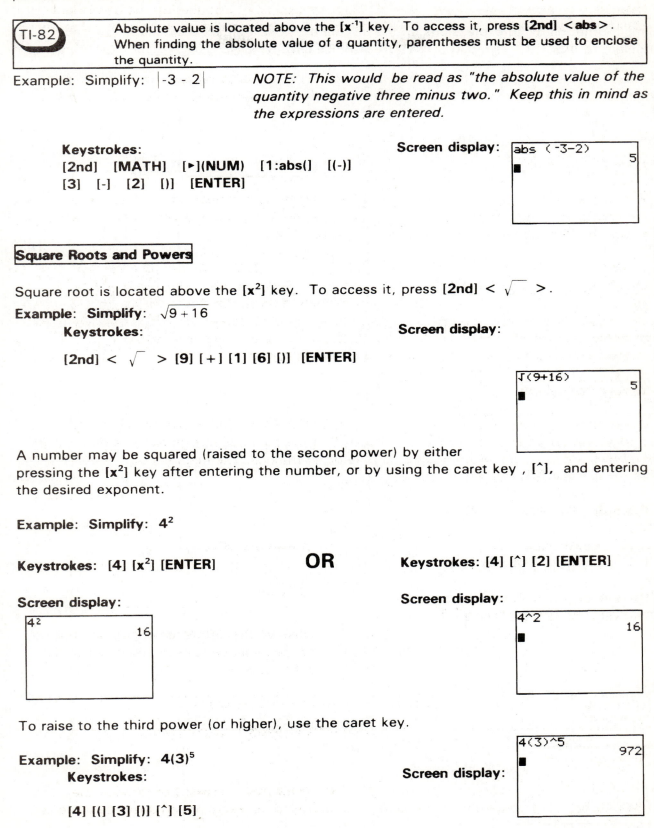

TI-82   Absolute value is located above the [x^{-1}] key. To access it, press **[2nd] <abs>**. When finding the absolute value of a quantity, parentheses must be used to enclose the quantity.

Example: Simplify: $|$-3 - 2$|$ *NOTE: This would be read as "the absolute value of the quantity negative three minus two." Keep this in mind as the expressions are entered.*

Keystrokes: **Screen display:**
[2nd] [MATH] [►](NUM) [1:abs(] [(-)]
[3] [-] [2] [)] [ENTER]

```
abs (-3-2)
                    5
■
```

Square Roots and Powers

Square root is located above the [x^2] key. To access it, press **[2nd] < √ >**.

Example: Simplify: $\sqrt{9+16}$

     Keystrokes: **Screen display:**

   [2nd] < √ > [9] [+] [1] [6] [)] [ENTER]

```
√(9+16)
                    5
■
```

A number may be squared (raised to the second power) by either pressing the [x^2] key after entering the number, or by using the caret key, [^], and entering the desired exponent.

Example: Simplify: 4^2

Keystrokes: [4] [x^2] [ENTER]   **OR**   **Keystrokes: [4] [^] [2] [ENTER]**

Screen display: **Screen display:**

```
4²
        16
```

```
4^2
        16
■
```

To raise to the third power (or higher), use the caret key.

Example: Simplify: $4(3)^5$
    Keystrokes: **Screen display:**

   [4] [(] [3] [)] [^] [5]

```
4(3)^5
        972
■
```

3

EXERCISE SET

Simplify the following expressions on the calculator.   Use the box provided below each problem to **RECORD THE CALCULATOR SCREEN <u>LINE BY LINE EXACTLY</u>** as it appears.

1. $10 \div 2 + 7 - 3$

2. $18 - (-3)$

3. $|-8|$

4. $-|-2|$

5. -3^4

6. $(-3)^4$

7. $|4^5 - (-6^2)|$

8. $|-4.7 - 5.28| - 18.3$

9. $-|-2^4 - 7|$

10. $\sqrt{5^3 - 10^2}$

11. $-\sqrt{169 - 25}$

12. $\sqrt{(-5)^2 - 4^2} + 16$

4

13. $(15 - 2)^3$

14. $|7 - 11| - \sqrt{64}$

✍ 15.   Explain the importance of using parentheses when entering expressions containing absolute value, roots, and exponents.

✍ 16.   Explain why entering - 8, negative eight, as "minus 8" (using the subtraction key) is incorrect. Explain the result this incorrect keystroking produces.

Shortcut Keys

Below are descriptions of some keys that may be helpful in the efficient use of the calculator. You may want to reference this material in the future.

[QUIT]

To return to the home screen press **[2nd] <QUIT>**. This is helpful when stuck on a screen and pressing **[CLEAR]** does not return the home screen.

TI-85/86	THE **[EXIT]** KEY ON THE TI-85/86 WILL REMOVE MENUS FROM THE SCREEN.

[INS]

This key is helpful when data is entered incorrectly - particularly in expressions that are lengthy. When using the insert key, first place the cursor in the position in which the inserted digit/symbol should appear, press **[2nd] <INS>** (the character under the cursor will blink) and then the desired digit/symbol(s) to be inserted. The calculator will insert as many characters as desired as long as a cursor key is not pressed.

[DEL]

This key is helpful when an incorrect key is mistakenly pressed. Place the cursor over the character to be deleted and press **[DEL]**.

[π]

When evaluating formulas requiring the use of π, press **[2nd]** $< \pi >$. Although only nine decimal places are displayed, the calculator will evaluate the expression using an eleven decimal place approximation for π.

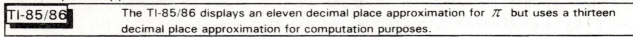

TI-85/86	The TI-85/86 displays an eleven decimal place approximation for π but uses a thirteen decimal place approximation for computation purposes.

[ENTRY]

Pressing **[2nd]** **<ENTRY>** accesses the ability of the calculator to recall the expression previously entered. Pressing **[2nd]** **<ENTRY>** repeatedly, performs "deep recall" by scrolling up the screen.

TI-85	The TI-85 does not have "deep recall." It will only recall the previous display line.

[2nd] [◄]

These keystrokes move the cursor to the beginning of the entry.

[2nd] [►]

These keystrokes move the cursor to the end of the entry.

[ANS]

The calculator will recall the answer from a previous computation. Access this function by pressing **[2nd]** **<ANS>**. ANS is located above the gray key used to enter a negative sign. It will also be activated if you press an operation key (+, -, X, ÷) before entering a number.

Solutions:

1. 9	2. 21	3. 8	4. -2	5. - 81	6. 81	7. 1060
8. - 8.32	9. - 23	10. 5	11. - 12	12. 19	13. 2197	14. - 4

15. Absolute value and root symbols are grouping symbols, therefore, parentheses are required to designate the grouped quantity. An exponent only applies to a single number/character unless grouping is used.

16. "Minus 8" indicates the operation of subtracting 8 from another quantity, whereas "negative 8" specifies the opposite of the number 8. Therefore, when a minus sign is used the calculator will retrieve the previous answer (ANS) for the 8 to be subtracted from.

Touring the TI-82

Take a few minutes to study the TI-82 graphing calculator. The keys are color-coded and positioned in a way that is user friendly. Notice there are dark blue, black, and gray keys, along with a single light blue key.

Dark blue keys: On the right side of the calculator are the dark blue keys. At the top are four directional cursor keys. These may be used to move the cursor on the screen in the direction of the arrow printed on the key. The four operation symbols (addition, subtraction, multiplication, and division) are also in dark blue. Notice the key marked **ENTER**. This will be used to activate commands that have been entered; thus there is no key on the face of the calculator with the equal sign printed on it.

Gray keys: The twelve gray keys that are clustered at the bottom center are used to enter digits, a decimal point, or a negative sign. Notice the gray key beneath the light blue key in the upper left position, labeled **ALPHA**. We will come back to this key shortly; because its position is different from the other gray keys, it serves a different function.

Black keys: The majority of the keys on the calculator are black. Below the screen are five black keys labeled [Y =], [**WINDOW**], [**ZOOM**], [**TRACE**], and [**GRAPH**]. These keys are positioned together below the screen because they are used for graphing functions. Notice the [X,T,θ] key in the second row and second column. It will be used frequently in algebra to enter the variable **X**. The **ON** key is the black key located in the bottom left position.

Light blue key: The only light blue key is the **2nd** key located in the upper-left position.

Above most of the keys are words and/or symbols printed in either light blue or white. To access a symbol in light blue (printed above any of the keys) first press the light blue [**2nd**] key, and then the key BELOW the symbol (function) to be accessed. For example, to turn the calculator **OFF** notice that the word **OFF** is printed in light blue above the **ON** key. Therefore, press the keys [**2nd**] [**ON**] to turn the calculator off. These keystrokes are done *sequentially* - not simultaneously. Throughout this book the following symbolism will be used: symbols that appear on the key will be denoted in brackets, [ ], whereas symbols written above the key will be denoted in < >. Thus, the previous command for turning the calculator off would appear as [**2nd**] <**OFF**>. The symbols [ ] or < > will cue you <u>where</u> to look for a command - either printed on a key or above it.

Alphabet letters and other symbols printed in white above some of the keys are accessed by first pressing the gray **ALPHA** key and then the key *below* the desired letter or symbol. Again, the keystrokes are sequential.

Note: The TI-82 has an Automatic Power Down (APD) feature which turns the calculator off when no keys have been pressed for several minutes. When this happens, press [ON] to access the last screen used.

Let's Get Started!

Turn the calculator on by pressing [**ON**]. If the display is not clear, press [**2nd**] [▲] to darken the screen, or [**2nd**] [▼] to lighten the screen. Notice that when the [**2nd**] key is pressed, an arrow pointing up appears on the blinking cursor.

To ensure the calculator is in the desired mode, press [**MODE**]. All of the options on the far left should be highlighted. If not, use the [▼] to place the blinking cursor on the appropriate entry and press [**ENTER**]. Exit MODE by pressing [**CLEAR**]. This accesses the home screen which is where expressions are entered. Press [**CLEAR**] until the screen is cleared except for the blinking cursor in the top left corner.

☞   Return to the core unit section entitled "Integer Operations" (pg.2) and complete the unit.

Touring the TI-85/86

Take a few minutes to study the TI-85/86 graphing calculator. The keys are color-coded and positioned in a way that is user friendly.

Gray keys: The twelve gray keys that are clustered at the bottom center are used to enter digits, a decimal point, and a negative sign. At the top right are four directional cursor keys. These may be used to move the cursor on the screen in the direction of the arrow.

Black keys: The majority of the keys on the calculator are black. The four operation symbols (division, multiplication, subtraction and addition) are located in the column on the far right. Notice the key marked **ENTER** at the bottom right position. This will be used to activate commands that have been entered. The equal sign on the face of the calculator is NOT used for computation. The key marked **x-VAR** in the second row, second column will be used frequently to enter the variable **x**. The **ON** key is in the bottom left position.

Below the screen are five black keys labeled **F1**, **F2**, **F3**, **F4** and **F5**. These are menu keys and will be addressed as they are needed.

Yellow-orange key: This key is located in the top-left position and is labeled **2nd**.

Blue key: This key is labeled **ALPHA** and is located at the top left of the key pad.

Above most of the keys are words and/or symbols printed in either yellow-orange or blue. To access a symbol in yellow-orange (printed above any of the keys), first press the yellow-orange [**2nd**] key, and then the key BELOW the symbol (function) you wish to access. For example, to turn the calculator **OFF** notice that the word **OFF** is printed in yellow-orange above the **ON** key. Therefore, press the **2nd** key and the **ON** key to turn the calculator off. These keystrokes are done _sequentially_ - not simultaneously. Throughout this book the following symbolism will be used: symbols that appear on the key will be denoted in brackets, [ ], whereas symbols written above the key will be denoted in < >. A function that is accessed from a menu will be written in parentheses, ( ). Thus, the previous command for turning the calculator off would appear as [**2nd**] <**OFF**>. The symbols [ ], < >, or ( ) will cue you _where_ to look for a command - either printed on a key, above it, or as a menu option. When a menu key is indicated, the current function of the key will follow in parentheses to correspond to the display at the bottom of the screen. For example, press [**GRAPH**] to display the graph menu. The notation [**F2**](**RANGE**) would denote access of the range submenu (if using the TI-86, [**F2**] (**WIND**) is displayed and denotes access to the WINDOW submenu). Users should be aware that the menu denoted as RANGE on the TI-85 and WIND on the TI-86 corresponds to the WINDOW menu referred to for the TI-82/83. To remove the menu at the bottom of the screen, press [**EXIT**].

To access a symbol printed in blue above some of the keys, first press the blue **ALPHA** key and then the key _below_ the desired letter or symbol. Again, the keystrokes are sequential. _NOTE: The **A**utomatic **P**ower **D**own (APD) feature turns the calculator off when no keys have been pressed for several minutes. When this happens, press [**ON**] to access the last screen used._

Let's Get Started!

Turn the calculator on by pressing **[ON]**. If the display is not clear, press **[2nd]** and hold down **[▲]** to darken the screen or **[2nd] [▼]** to lighten. Notice that when the **[2nd]** key is pressed an arrow pointing up appears on the blinking cursor.

To ensure the calculator is in the desired mode, press **[2nd] <MODE>**. All of the options on the far left should be highlighted. If not, use the **[▼]** to place the blinking cursor on the appropriate entry and press **[ENTER]**. Exit **MODE** by pressing **[EXIT]**, **[CLEAR]**, or **[2nd] <QUIT>**. This accesses the home screen where expressions are entered.

Integer Operations

When entering integers on the calculator, differentiation must be made between a subtraction sign and a negative sign. Notice the subtraction sign is in the right column whereas the negative sign appears as a gray key next to the **[ENTER]** key and is labeled **(-)**.

☞ STUDY THE EXAMPLE IN THE SECTION "INTEGER OPERATIONS" IN THE CORE UNIT (PG. 2) AND CONTINUE UNTIL THE NEXT TI-85/86 PROMPT.

Absolute Value

The TI-85/86 does not have absolute value on its keypad. However, a key can be created using the **[CUSTOM]** key. Up to fifteen frequently used functions can be customized (and accessed with only two key strokes) provided the function desired is listed under **CATALOG**. To customize a function, press **[2nd] <CATALOG>** (if using the TI-86, press **[F1] (CATLG)** next) followed by **[F3](CUSTM)**. Place the arrow next to **abs** in the list, then press **[F1]**. The function **abs** is now listed under **PAGE↓**. The same procedure can be used to add other functions to the **CUSTOM** menu as necessary. Each time select an open slot in the menu by choosing the menu key below the open slot. Pressing **[MORE]** accesses additional slots for customizing. When finished, press **[EXIT]** until the menus at the bottom of the screen clear, and the home screen is displayed.

Pressing **[CUSTOM]** reveals the customized menu; to access a customized function, simply press the **F** key below the desired function. Pressing **[EXIT]** removes this menu.

Example: Simplify | -3 - 2 |
(Ensure you are at the home screen - press **[EXIT]** if necessary.)

Keystrokes: **Screen display:**
[CUSTOM] [F1](abs) [(] [(-)] [3] [-]
[2] [)] [ENTER]
Note: Absolute value can also be accessed by pressing **[2nd] <MATH> [F1](NUM) [F5](abs)**.

☞ RETURN TO THE "ABSOLUTE VALUE" SECTION OF THE CORE UNIT (PG. 2) AND CONTINUE UNTIL THE NEXT TI-85/86 PROMPT. KEYSTROKES FOR YOUR CALCULATOR WILL DIFFER FROM THOSE GIVEN FOR THE TI-82/83/83PLUS; MAKE APPROPRIATE CHANGES IN THE MARGIN.

UNIT 2
FRACTIONS AND THE ORDER OF OPERATIONS

*Unit 1 is a prerequisite for this unit.

TI-85/86 | TI-85/86 USERS TURN TO "FRACTIONS AND THE TI-85/86" IN THE GUIDELINES (PG.17).

Fractions and the TI-82/83/83plus

The calculator can be used to perform arithmetic operations with fractions. Pressing the [**MATH**] key reveals a math menu. Take a minute to use the [▼] cursor to scroll down the menu. Notice that there are 10 options available. Use of the first option, **1:▸Frac**, will now be illustrated.

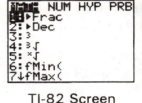

TI-82 Screen   TI-83/83plus Screen

Enter the decimal number 0.42 by pressing [.] [4] [2] [MATH]. Under the MATH menu **1:▸Frac** is highlighted. Because it is highlighted, press [**ENTER**] to select option number **1**. Press [**ENTER**] again to activate the "convert to a fraction" command. Notice that the calculator displays the fraction reduced to lowest terms.

```
.42▸Frac
            21/50
```

Recall that a mixed number is actually the sum of an integer and a fraction. Therefore, to enter mixed numbers on the TI-82/83/83plus simply indicate the addition of the integer and the fraction. To enter
$-3\frac{1}{2}$, press [(-)] [3] [+] [(-)] [1] [÷] [2] [MATH] [ENTER] (to

```
-3+ -1/2▸Frac
            -7/2
```

select option number **1**) [**ENTER**]. Your calculator display should correspond to the one at the right. Because the entire fraction is negative, both the integer part and the rational part must be entered as negative numbers.

Example 1: Simplify.

Keystrokes: $\frac{2}{3} + \frac{1}{5}$ Screen display:

```
2/3+1/5▸Frac
            13/15
```

[2] [÷] [3] [+] [1] [÷] [5] [MATH] [ENTER] [ENTER]

NOTE: If a denominator is more than four digits, the decimal equivalent of the fraction will be returned. Enter the fraction $\frac{17}{10,000}$ **and activate the convert to Frac option. The calculator will only display the decimal form of the number.** ◆

Order of Operations

Recall the order of operations: a. Simplify expressions within grouping symbols and exponents.

b. Perform the operations of multiplication and division from left to right.

c. Perform the operations of addition and subtraction from left to right.

The calculator is programmed to follow this order of operations. Therefore, the calculator performs the operations of multiplication and division before the operations of addition and subtraction. Moreover, it does the multiplication and division in the order in which they appear from left to right. It then performs the additions and subtractions in the order in which they appear from left to right.

To ensure understanding of this order of operations, compare the following screen displays for the addition of two fractions:

Example 2: Simplify $\dfrac{5}{8} + \dfrac{2}{3}$.

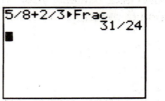

```
5/8+2/3►Frac
            31/24
■
```

```
(5/8)+(2/3)►Frac
              31/24
■
```

The calculator automatically performs the division before the addition; thus, insertion of the parentheses is not necessary.   ◆

However, when dividing two fractions, care must be taken to insert parentheses correctly. Consider the following screen displays for the division of two fractions:

Example 3: Simplify $\dfrac{5}{7} \div \dfrac{3}{4}$.

```
(5/7)/(3/4)►Frac
          20/21
■
```

```
5/7/3/4►Frac
            5/84
■
```

The screen on the left is correct. The screen on the right indicates that the division is performed from left to right, producing the expression $\dfrac{5}{7} \div 3 \div 4$, which is not equivalent to the original expression.   ◆

Notice that the order of operations dictates that operations enclosed within grouping symbols must be performed first. There are six grouping symbols used in mathematics:

parentheses: ( )   brackets: [ ]   braces: { }

absolute value: | | root symbol: $\sqrt{}$ fraction bar: ─────

ONLY parentheses are used as grouping symbols on the grapher. Even though brackets and braces appear on the calculator face, they are *not* programmed for grouping.

Example 4: Simplify 5{8 + [3(4 - 2)]}

If brackets,[ ], and braces, { }, were entered, you will get an error message. Moreover, the calculator displays it as a SYNTAX error. This means that a symbol was entered inappropriately.

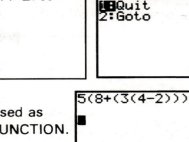

On the graphing calculator, brackets and braces *are not* used as grouping symbols. ONLY PARENTHESES SERVE THAT FUNCTION.

Before pressing **[ENTER]**, it is a good idea to count the number of parentheses entered. There should be an even number. Insert any that were inadvertently omitted.

◆

Explorations Unit 1 illustrated the use of parentheses with the square root and absolute value symbols. Go back to this unit and refresh your memory on the use of parentheses when using these symbols. The next example illustrates the use of parentheses with the fraction bar.

Example 5: Simplify $\dfrac{8 + 16}{2}$

The numerator must be enclosed in parentheses to ensure that the entire quantity is divided by 2. Any time a numerator and/or a denominator contain more than one character, parentheses must be used.

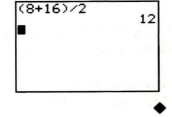

◆

EXERCISE SET

Simplify the following expressions on the calculator; all answers should be expressed as fractions. **RECORD THE CALCULATOR SCREEN <u>LINE BY LINE EXACTLY</u>** as it appears.

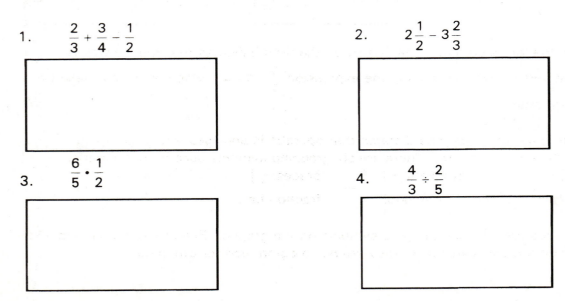

1. $\dfrac{2}{3} + \dfrac{3}{4} - \dfrac{1}{2}$

2. $2\dfrac{1}{2} - 3\dfrac{2}{3}$

3. $\dfrac{6}{5} \cdot \dfrac{1}{2}$

4. $\dfrac{4}{3} \div \dfrac{2}{5}$

5. $\left| -\dfrac{3}{5} \right| \cdot \sqrt{\dfrac{1}{4}}$

6. $\left(\dfrac{1}{2} \right)^3 \cdot \dfrac{2}{3}$

7. $\sqrt{\dfrac{4}{25}} \div 2$

8. $\left(\dfrac{3}{5} \right)^2 \cdot \left(\dfrac{2}{3} \right)^3$

9. $\dfrac{3}{5} \div \dfrac{1}{2} - \dfrac{5}{8}$

10. $\dfrac{5 - |1 - 3| + 7}{6 - 9 \div 3}$

11. $\dfrac{3 - |-15| \div 5}{-3(5 - |-2|) + 15}$

12. $\dfrac{6 + 12 \div 2 \cdot 3 - 5}{3^2 - 5^2 \cdot 3 + 6 \cdot 11}$

Note: Be sure to translate any error messages to the appropriate mathematical notation.

✍ 13.   Enter the expression $\dfrac{16 + 8}{4}$ on your calculator in each of the following ways:

16 + 8 / 4 and (16 + 8)/4. Which is the correct format and why?

✏ 14.   IN YOUR OWN WORDS: Summarize what you learned in this unit about entering fractions on the calculator and the order of operations. Go back through the unit and look at the displays copied for problems as well as written responses to questions. Include in your summary the use of grouping symbols such as {}, [], square root symbols and absolute value bars.

Solutions:

1. 11/12   **2.** -7/6   **3.** 3/5   **4.** 10/3   **5.** 3/10   **6.** 1/12   **7.** 1/5

8. 8/75   **9.** 23/40   **10.** 10/3   **11.** 0   **12.** undefined

13. The numerator must be grouped in parentheses, (16 + 8), in order for the calculator to divide the denominator, 4, into the entire numerator.

Fractions and the TI-85/86

Since the ▶**Frac** option is used frequently, it will be added to the CUSTOM menu. To do this, press **[2nd]** **<CATALOG>** **[F3](CUSTM)**. Cursor down the list to place the arrow next to ▶**Frac**, found near the bottom of the list. Press **[F2]** or another open **F** key to create the custom key. HINT: Since ▶**Frac** was found at the bottom of the list, pressing **[▲]** **[2nd]** **<M2>(PAGE↑)** until the arrow is positioned next to the desired selection would be quicker than scrolling through the entire list.

Note: To delete a customized entry, press **[2nd]** **[CUSTOM]** **[F4](BLANK)** and then the **F** key that contains the command to be deleted.

Press **[EXIT]** until the blinking cursor is displayed at the HOME screen.

Enter the decimal number 0.42 by pressing **[.]** **[4]** **[2]** **[CUSTOM]** **[F2](▶Frac)** **[ENTER]**. Notice that the calculator displays the fraction reduced to lowest terms.

Recall that a mixed number is actually the sum of an integer and a fraction. Therefore, to enter a mixed number, simply enter the expression as an indicated sum. To enter $-3\frac{1}{2}$ and express it as a fraction, press **[(-)]** **[3]** **[+]** **[(-)]** **[1]** **[÷]** **[2]** **[CUSTOM]** **[F2](▶Frac)**. Because the entire fraction is negative, both the integer part and the rational part must be entered as negative numbers. Your display should correspond to the one at the right.

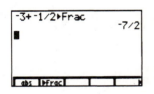

☞ Return to Example 1 in the section "Fractions and the TI-82/83/83plus" of the core unit (pg.11) and complete the unit.

UNIT 3
THE STOre AND TABLE FEATURES

*Unit 2 is a prerequisite for this unit. Answers appear at the end of this unit.

Evaluating an expression means to find the value of the expression for given values of the variables. To analytically evaluate expressions, the substitution property is applied when constant values are substituted for specified variables. The expression is then simplified following the order of operations. The STOre feature and the TABLE of the calculator can be used to accomplish the same result.

Using the STOre Key

 TI-85/86   IF USING THE TI-85/86, GO TO THE GUIDELINES (PG. 27).

> NOTE: Because **X** is used as a variable so often in algebra, it has its own key on the calculator. The calculator is in function MODE; therefore when [**X,T,θ,n**] is pressed the screen displays the variable **X**.
>
> TI-82   **This key appears as [X,T,θ] on the TI-82.**

To store a value, simply enter the value, press [**STO▸**] and then the desired variable. For example, to store X = 5, press [**5**] [**STO▸**] [**X,T,θ,n**] [**ENTER**]. The value 5 is now stored under the variable X. This value will remain stored in X until another value replaces it. To check and see what value is actually stored under a specific variable, enter the variable at the prompt (blinking cursor) and then press [**ENTER**]. The value is then displayed on the screen.

Example 1:   Evaluate $3X^2 + 6X + 2$ when X = -11.

Solution: Store X = -11, by pressing [**(-)**] [**1**] [**1**] [**STO▸**] [**X,T,θ,n**]. The colon ":" key (located above the gray decimal key) is used to separate commands that are to be entered on the same line, so now press [**ALPHA**] <:> before entering the expression to be evaluated.

> TI-82   **TI-82 users should press [2nd]< : >.**

Press [**3**] [**X,T,θ, n**] [**x²**] [**+**] [**6**] [**X,T,θ,n**] [**+**] [**2**] to enter the expression. At this point, the calculator has been instructed to store -11 in X and then evaluate $3X^2 + 6X + 2$ for this value of X. The calculator will not perform the instructions until [**ENTER**] is pressed. The polynomial evaluates to 299. Your screen should look like the one at the right.

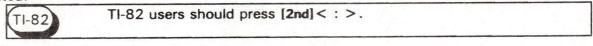

To store a value under a variable other than **X**, the [**ALPHA**] key must be used to access the letters of the alphabet. Pressing the [**ALPHA**] key, followed by another key, allows access to the upper case letters and symbols written above the key pads in the same color as the alpha key.

19

Example 2:   Evaluate $a^2 - 3a + 5$ when $a = -7$.

```
-7→A:A²-3A+5
                 75
■
```

Solution: See screen at right.

NOTE: Be sure to use the [(-)] key for the negative sign on a number and use the subtraction key, [-], to indicate subtraction.

◆

EXERCISE SET

✍1.   Look at the screen displayed at the right. What have you told the calculator to do?

```
-2→X:3→Y:X²-2X+Y
2
                 17
```

   a. -2→X means _____

   b. 3→Y means _____

   c. $X^2 - 2X + Y^2$ means _____

   d. 17 means _____

   e. The purpose of the colon (:) is to_____

Exercises 2-6 should be completed as follows:
(a) evaluate each expression analytically by substituting the values for the variables into the given expression and following the order of operations, and
(b) evaluate with the calculator, using the **STO►** feature. Record the screen display as a means of showing your work. You must copy the screen display <u>exactly</u> as it appears, line by line.

2.   Evaluate $2X^2 + 3X + 1$ when $X = -3$.

   Algebraically: *Calculator display:*

3.   Evaluate $(X - 2)^2 + 3X$ when $X = \frac{1}{2}$.

   Algebraically: *Calculator display:*

20

4.   Evaluate 4(X - 3) + 5(Y - 3) - 4 when X = -1 and Y = -3.

Algebraically: *Calculator display:*

5.   Evaluate $-X^2 + 3XY^2 - 6Y^3$ when X = 3 and Y = 4.

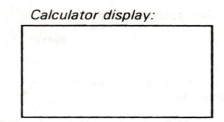
TI-85/86   IF USING THE TI-85/86, GO TO THE GUIDELINES (PG.28).

Algebraically: *Calculator display:*

6.   Evaluate $\dfrac{2X^3Y - 2X^2Y + X}{5X - Y}$ when X = 2 and Y = ½.

Algebraically: *Calculator display:*

7.   Troubleshooting: Each of the problems below has been entered **incorrectly** on the calculator. Use the STOre feature to evaluate the expression on your calculator and copy the screen display exactly as it appears. Then circle the displayed error.

a. Evaluate 2X - 5 when X = -3.

X→ -3: 2X-5▮

b. Evaluate $\dfrac{\sqrt{B^2 - 4AC}}{2A}$ when B = 4, A = 5, C = -1.

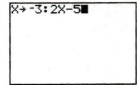
5→A: 4→B: -1→C: √(B
²-4AC)/2A

✍ 8. Discuss the use of parentheses when entering expressions containing radicals or fractions.

Checking Solutions

The graphing calculator can be used to check solutions to equations and to confirm that expressions are equivalent.

Example 3: Is 6 a solution to the equation $4(2X - 1) - 8 = 42 - X$?

Solution: If 6 is a solution to the equation then the left side, $4(2X - 1) - 8$, will have the same value as the right side, $42 - X$, when 6 is substituted for X. Store 6 in X and then find the value of each side of the equation. Your screen should look like the one at the right.

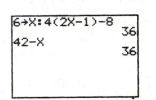

◆

TI-85 BECAUSE THE TI-85 DOES NOT HAVE A TABLE FEATURE, TI-85 USERS SHOULD USE THE STO► KEY TO DUPLICATE TABLE RESULTS IN THE FOLLOWING EXAMPLES. YOU MAY WANT TO EXPLORE THE WWW.TI.COM/CALC/DOCS/ARCH.HTM WEB SITE FOR TABLE PROGRAMS FOR YOUR CALCULATOR.

A proposed solution can also be verified through the TABLE feature of the TI-82, TI-83/83plus, or TI-86 graphing calculators.

Example 4: Consider the equation $3(5x + 1) = 8x + 24$. Is 3 a solution to this equation?

Solution: To confirm that 3 is a solution via the TABLE, we will *store* the expressions on each side of the equation at the **y-edit** screen. To do this, press [Y=] and CLEAR all entries.

TI-86 PRESS [GRAPH] [F1](Y(X)=).

Enter the left side of the equation at the **Y1=** prompt. Pressing [ENTER] will move the cursor to the **Y2=** prompt, which is where the right side of the equation is to be entered. To display the appropriate TABLE value, press [2nd] <**TblSet**>.

TI-86 TI-86 USERS PRESS [TABLE] [F2](TBLST).

The cursor is now at the **TblStart=** prompt (TblMin= on the TI-82). Enter 3, since that is the value to be verified as a solution. Press [ENTER] to move the cursor to the **ΔTbl=** prompt. Enter a 1 (one) to increment the X- value of the TABLE by one unit. (Since we will only be concerned with one entry line in the TABLE, the table increment is not critical. It is suggested that the increment be left at 1.) Access the TABLE by pressing [2nd] [TABLE].

TI-86 TI-86 USERS PRESS [F1](TABLE).

Your screen should look like the one pictured to the right. Notice the first entry under the X column is 3. Recall the expression $3(5x + 1)$ was entered at the Y1= prompt and $8x + 24$ was entered at the Y2= prompt. The values in the Y1 and Y2 column are equal (i.e. both 48) verifying that when $x = 3$, each of the expressions entered at the Y1= and Y2= prompts evaluate to 48.

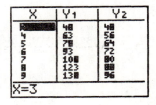

Use the [▲] and [▼] cursors to scroll through the TABLE. Notice that the values in the Y1 and Y2 columns are equal ONLY when x = 3.

The TABLE has provided numerical evidence that 3 is a solution to the equation $3(5x + 1) = 8x + 24$. In the space below, use the Substitution Property to evaluate both sides of the equation analytically (by hand) to confirm that x = 3 is a solution.

The **STO►** feature can also be used to confirm that when 3 is substituted into the expression on each side of the equation, the expressions both evaluate to 48.

STEPS FOR CONFIRMING A SOLUTION VIA THE TABLE

1. Press **[Y =]** and enter the left side of the equation at **Y1 =** and the right side at **Y2 =**.

    | TI-86 | TI-86 USERS PRESS **[TABLE] [F2](TBLST)**.

2. Press **[2nd] < TblSet >** to set the TblStart (TblMin on the TI-82) to the value believed to be the solution of the equation. The value for **ΔTbl** may be any value. It is suggested that the number one be used.

3. Access the TABLE by pressing [2nd] < TABLE > (TI-86 users press **[F1](TABLE)**.) The solution is confirmed if and only if the Y1 and Y2 columns are identical for the designated value of X.

EXERCISE SET

Directions: Verify that each number below is a solution to the given equation. Copy the line(s) of the TABLE that verify the solution(s). TI-85 users use the STOre feature to fill in the entries.

9. Verify that 3/2 is a solution to the equation $X + 5 = 3X + 2$

X	Y1	Y2

10. Verify that 2 is a solution to the equation $\frac{3}{2}(X + 4) = \frac{20 - X}{2}$

X	Y1	Y2

23

11. Verify that 4 **and** -3 are both solutions to the equation $X^2 - X - 12 = 0$.

X	Y1	Y2

12. What two values of X can be verified as solutions to an equation from the displayed table?

X = _____ and X = _____

✍ What assures you that these numbers are solutions?

X	Y1	Y2
-2	4	-1
-1	1	1
0	0	3
1	1	5
2	4	7
3	9	9
4	16	11

X=-2

✍13. The solutions to the equation $3X^2 = 5X + 12$ are -4/3 and 3. When a student attempts to confirm the solution of -4/3, he instructs the calculator to display the TblStart as -4/3 and to be incremented by 1. Note that the TblStart of -4/3 is converted to a decimal approximation for display in the TABLE. The top row of the TABLE confirms -4/3 as a solution. How can the solution X = 3 be confirmed?

X	Y1	Y2
-1.333	5.3333	5.3333
-.3333	.33333	10.333
.66667	1.3333	15.333
1.6667	8.3333	20.333
2.6667	21.333	25.333
3.6667	40.333	30.333
4.6667	65.333	35.333

X=-1.33333333333

Example 5: Simplify the following expression algebraically: $3X(X - 2) + 5X^2$; use the TABLE to confirm that your simplified expression is equivalent to the original expression.

Solution: The simplified expression is $8X^2 - 6X$. To determine if $3X(X - 2) + 5X^2$ equals $8X^2 - 6X$, enter the original expression at the **Y1 =**

```
TABLE SETUP
 TblStart=-4/3
 △Tbl=1
Indpnt: AUTO  Ask
Depend: AUTO  Ask
```

X	Y1	Y2
0	0	0
1	2	2
2	20	20
3	54	54
4	104	104
5	170	170
6	252	252

X=0

prompt and the simplified expression at the **Y2 =** prompt. Set the TABLE to have a start value of 0 and increment the TABLE by 1. This start value and the increment are arbitrary. Any values could be used in either position. Your screen should match the one displayed.

WARNING: This procedure confirms that two expressions are equivalent. It does not determine that the expression is completely simplified. ◆

14. Explain **how**, in Example 5, the use of the TABLE confirms that the given expressions are equal.

15. The expression $(3X^2 - 5X + 2) - (8X^2 - 3X - 1)$ simplifies to $-5X^2 - 2X + 3$. Confirm that these two expressions are equivalent by using the TABLE feature of the graphing calculator.
Describe the entries on the TABLE screen.

16. A student entered an expression at the **Y1=** prompt and what he believed to be the simplified form of the expression at the **Y2=** prompt. His screen display is pictured.

a. Are his expressions equivalent?

b. Why or why not?

X	Y₁	Y₂
	3	2
1	8	7
2	23	22
3	48	47
4	83	82
5	128	127
6	183	182

X=0

17. IN YOUR OWN WORDS: Write a summary of what you have learned in this unit. Look back at the screens you recorded and compare them to your hand-written arithmetic/algebraic expressions. Your summary should include:
a. The use of the [X,T,θ,n] key,
b. use of the colon on a command line,
c. evaluating expressions with the STO► key,
d. use of the TABLE feature to check solutions/roots of equations (TI-85 users should address the use of the STOre key to construct a TABLE),
e. use of the TABLE feature to verify that two expressions are equal.

<u>Solutions</u>:

1. a. store -2 for the variable x **b.** store 3 for the variable y **c.** evaluate the expression for the stored value of the variables **d.** the value of the expression is 17 when x = -2 and y = 3

e. separate commands

2. 10 **3.** 15/4 **4.** -50 **5.** -249 **6.** 12/19

7. a. Your screen display should indicate that -3 is stored in X: -3→X

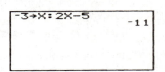

b. Always check that correct values for the variable have been stored:

8. answers may vary

9.

X	Y1	Y2
1.5	6.5	6.5

10.

X	Y1	Y2
2	9	9

11.

X	Y1	Y2
-3	0	0
4	0	0

12. X = -1, X = 3; The Y columns are equal for these X values.

13 Change the table minimum (start) to 3.

14. answers may vary

15. Entry by entry, all values of Y1 equal all values of Y2.

16. a. No **b.** They are not equivalent because the values in the Y1 column do not equal the values in the Y2 column.

26

The [x-VAR] key is used to display the variable **x**. Note that the variable **x** is displayed as a lowercase letter on the TI-85/86, instead of an uppercase **X** as the [X,T,θ,*n*] key on the TI-82/83/83plus. When using this text, each **X** printed as a variable for the TI-82/83/83plus should appear as **x** on the TI-85/86.

To store a value, simply enter the value, press [STO▸] and then the desired variable. For example, to store X = 5, press [5] [STO▸] [x-VAR] [ENTER]. The value 5 is now stored under the variable **x**. This value will remain stored in **x** until another value replaces it. To determine the value that is actually stored under a specific variable, enter the variable at the prompt (blinking cursor) and then press [ENTER]. The value is then displayed on the screen.

NOTE: Pressing the **STO▸** key automatically initiates the ALPHA key, allowing access to the remaining letters of the alphabet. Press **STO▸** and observe that the blinking cursor has an **A** in it for **ALPHA**. Letters will automatically be recorded in uppercase format. To disengage the ALPHA feature, press the [**ALPHA**] key.

Example 1: Evaluate $3X^2 + 6X + 2$ when X = -11.

Solution: Store X = -11, by pressing [(-)] [1] [1] [STO▸] [x-VAR]. The colon ":" key (located above the gray decimal key) is used to separate commands that are to be entered on the same line. Press [2nd] <:> before entering the expression to be evaluated. At this point you should note that the blinking cursor has an **A** in it for ALPHA. Disengage the ALPHA feature by pressing [**ALPHA**] now. Press [3] [x-VAR] [x²] [+] [6] [x-VAR] [+] [2] to enter the expression that is to be evaluated. At this point, the calculator has been instructed to store -11 in **x** and to evaluate $3x^2 + 6x + 2$ at this value of **x**. The calculator will not perform the instructions until the [**ENTER**] key is pressed. The polynomial evaluates to 299. Your screen should look like the one above.

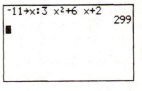

The TI-85/86 has both uppercase and lowercase alphabet letters that can be used as variables. To access a lowercase letter when using the **STO▸** feature, press [2nd] <ALPHA> after pressing the [STO▸] key. However, realize that an expression containing a lower case variable could be entered as an expression using an upper case variable as long as you are consistent. Use lower case letters with care. As noted at the end of this guideline, some lower case letters are reserved for system variables.

When not using the **STO▸** feature, uppercase letters are accessed by [**ALPHA**] and lowercase letters are accessed by [**2nd**] <ALPHA>.

Example 2: Evaluate $a^2 - 3a + 5$ when a = -7.

Solution: See screen at right.

NOTE: Be sure to use the gray [(-)] key for the negative sign on a number and use the black [-] key to indicate subtraction.

The keystrokes are [(-)] [7] [STO▸] [A] [2nd] <:> [A]*(observe that the cursor is still blinking with an uppercase A inside of it)* [ALPHA] *(the ALPHA feature is now disengaged)* [x²] [-] [3] [ALPHA] [A] [+] [5] [ENTER].

◆

☞   RETURN TO THE EXERCISE SET IN THE CORE UNIT (PG. 20) AND WORK THE EXERCISES.

The TI-85/86 has the capability of recognizing combinations of letters as independent variables. For example, the expression **4AC** means "four times A times C". However, the TI-85/86 would recognize **AC** as representing only <u>one</u> unknown, not a product of two unknowns. Thus the expression "four times A times C" must be displayed on the TI-85/86 as **4A*C** or **4A C**. Variables can be separated with a space [⌴] to denote the multiplication of two different quantities. By being able to combine letters to form new variables, the TI-85/86 has a limitless list of variables available for use. Uppercase letters are reserved for user variable names. Note that the TI-85/86 would recognize **4a*c** as an entirely different product since the variables are lowercase. Thus Exercise #5 on page 21 should be entered with either a multiplication sign or a **space** between the two variables in the middle term; i.e. $3X * Y^2$ or $3X\ Y^2$, but not $3XY^2$.

Lowercase letters should be used with care as some are reserved for system variables. Examples of these are:

a	coefficient of regression
b	coefficient of regression
c	speed of light
e	e natural log base
g	force of gravity
h	Planck's constant
k	Boltzman's constant
n	number of items in a sample
u	atomic mass unit

The variables $r, t, x, y,$ and θ are updated by graph coordinates based on the graphing mode.

☞   RETURN TO THE CORE UNIT EXERCISE 5 (PG. 21) AND COMPLETE THE UNIT.

UNIT 4
SCIENTIFIC NOTATION

*Unit 2 is a prerequisite for this unit. Answers appear at the end of the unit. This unit develops the concept of scientific notation through the process of exploration.

Scientific notation is an important application of integral exponents. It provides a means of writing very large or very small numbers in a compact form.

1. Use the graphing calculator to simplify each of the expressions below. The " $\times$ " used is a multiplication sign - *not* a variable.

    a. $2.56 \times 10^2 = $ _____

    b. $3.5 \times 10^3 = $ _____

    c. $6.2 \times 10^4 = $ _____

    d. $8 \times 10^5 = $ _____

    Compare the placement of the decimal in the original number to its placement in the value of the expression entered on the given lines. Notice how the decimal always moves to the right. Now compare the number of places the decimal moves to the power of 10 specified in the original expression. What is the relationship between the number of decimal moves and the power of 10?

2. Use the graphing calculator to again simplify each of the expressions below. Again, the " $\times$ " is used is a multiplication sign and *not* as a variable.

    a. $3.658 \times 10^{-1} = $ _____

    b. $2 \times 10^{-2} = $ _____

    c. $7.2 \times 10^{-3} = $ _____

    Compare the placement of the decimal in the original number to its placement in the value of the expression. Notice how the decimal always moved to the left. Now compare the number of places the decimal moved to the power of 10 specified in the original expression. Describe the relationship between the negative power of ten and the decimal moves.

29

3. Based on the conclusions drawn from #1 and #2, simplify each of the following expressions *without* using the calculator. Use the patterns observed in the first two problems.

a. $4.65 \times 10^4 =$ _____

b. $6.3 \times 10^3 =$ _____

c. $7.658 \times 10^5 =$ _____

d. $6.2 \times 10^{-2} =$ _____

e. $9.5 \times 10^{-5} =$ _____

Check the results recorded above by simplifying the expressions using the calculator.

Each of the numbers above is written in scientific notation. This is a notation used by scientists to write very large or very small numbers. When writing a number in scientific notation, the first number is ALWAYS greater than or equal to 1 and less than 10. The power of ten indicates the number of places the decimal is moved - movement is to the left when the exponent is negative, to the right when the exponent is positive. In scientific notation it is customary to use the "$\times$" to indicate multiplication rather than parentheses or a dot.

---

4. The distance that light travels in one year is approximately 9,460,000,000,000 km. If expressed as 94.6×10^{11} it **IS** written as the product of a number and a power of 10, but it **IS NOT** written in scientific notation. Why not?

Rewrite the number correctly in scientific notation: _____

5. Write each of the numbers below in scientific notation:

a. 6,000 _____
(the temperature of the sun in degrees Celsius)

b. 380,000,000 _____
(the distance from the earth to the moon in kilometers)

c. 0.000 000 022 _____
(the diameter of a helium atom in centimeters)

d. 0.000 000 000 000 000 000 000 001 67 _____
(the gram weight of a hydrogen atom)

6. The graphing calculator has a feature that allows all computation to be expressed in scientific notation. After turning the calculator on, press [**MODE**]. (TI-85/86 PRESS [**2ND**] <**MODE**>.) All of the options on the left should be highlighted. Change the setting on the first line from **NORMAL** to **SCI** (scientific) by using the right cursor arrow to place the blinking cursor over **SCI**. To select this mode, press [**ENTER**].

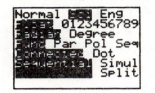

TI-82 Screen

TI-83/83plus Screen

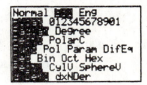

TI-85/86 Screen

Press [**CLEAR**] to return to the home screen.

7. Now enter the numbers below (one at a time) into the calculator. Press [**ENTER**] after each entry. Copy the displayed result.

   a. 6,000 _____

   b. 380,000,000 _____

   c. 0.000 000 022 _____

   d. 0.000 000 000 000 000 000 000 001 67 _____

8. In #5a, 6,000 should have been written as 6×10^3 in scientific notation. However, in #7a, the calculator displayed that number as **6 E 3**. The notation the calculator displays differs from the notation we use. Explain/reconcile the difference.

9. Complete the chart below by entering the number on the left in the calculator, recording the display, and then writing the number in scientific notation. The first one has been done to provide a model.

Standard Notation	Calculator Display	Scientific Notation
8,000	8 E 3	8×10^3
a. 0.00358		
b. 2,000,000		
c. 0.0124		
d. 67,300		

10. Since the calculator will convert very large or very small numbers to scientific notation, it may sometimes prove easier to enter very large or very small numbers into the calculator using the calculator's method of display. For example, the weight of the hydrogen atom in 5*d* is 0.000 000 000 000 000 000 000 001 67 grams. In scientific notation this is 1.67×10^{-24} which is displayed on the calculator as 1.67 E -24. When using this number in a computation it would be quicker to enter it as 1.67 E -24. The E notation can be displayed by pressing **[2nd] [EE]**.

11. Multiply $(5 \times 10^3)(3 \times 10^4)$. Use the EE key when entering the problem on the calculator. Record your screen display:

    a. Convert the displayed result on the calculator to scientific notation:_____

    b. Express the result in standard notation:_____.

12. Complete each computation below by entering each expression on the calculator and pressing **[ENTER]**. Expressions should be entered as displayed. Record the calculator display beneath the problem as a means of showing your work. Record the result in scientific notation on the blank provided.

    a. $(0.0006)(10^{-7})$ _____

    b. $\dfrac{(6,000,000)(40,000)}{3,000}$ _____

    c. $\dfrac{(0.0000008)(5,000,000)}{(0.0004)(0.00005)}$ _____

13. Convert the answers above to standard notation by either using the rules stated previously in #1 and #2 and/or by resetting the calculator back to **NORMAL** mode. To do this, press **[MODE] [ENTER]**. Return to the home screen by pressing **[CLEAR]**.

    a._____   b._____   c._____

14. Even when **NOT** in **SCI** mode, the calculator will express very large or very small numbers in scientific notation. For example: With the calculator in **NORMAL MODE**, simplify the following expression. Record the calculator display in the space below:

(0.00006)(0.0000008)

   a. Write the answer in scientific notation (not calculator notation!) _____

   b. Write the answer in standard notation: _____

✍15. IN YOUR OWN WORDS: Summarize what you learned in this unit. Your summary should address:
   a. converting a number from standard notation to scientific notation (both with the calculator and without it),
   b. converting a number from scientific notation to standard notation (both with the calculator and without it), and
   c. use of the EE key on the calculator.

Solutions: **1a.** 256   **1b.** 3500   **1c.** 62000   **1d.** 800000   **2a.** 0.3658   **2b.** 0.02

2c. 0.0072   **3a.** 46500   **3b.** 6300   **3c.** 765800   **3d.** 0.062   **3e.** 0.000095

4. The integer part (94) is greater than 10. 9.46×10^{12} **5a.** 6×10^{3} **5b.** 3.8×10^{8}

5c. 2.2×10^{8} **5d.** 1.67×10^{-24} **7a.** 6 E 3 **7b.** 3.8 E 8   **7c.** 2.2 E -8

7d. 1.67 E -24 **8.** "E" followed by a number indicates "times 10 to the power of the

number that follows"

9.

.00358	3.58 E-3	3.58×10^{-3}
2,000,000	2 E6	2×10^{6}
.0124	1.24 E-2	1.24×10^{-2}
67,300	6.73 E4	6.73×10^{4}

10.

2.15922×10^{-2}

8×10^{7}

2×10^{8}

11a. 1.5×10^{8} **11b.** 150,000,000   **12a.** 6×10^{-11} **12b.** 8×10^{7} **12c.** 2×10^{8}

14a. 4.8×10^{-11} **14b.** 0.000 000 000 048

```
┌─────────────────────────────────────────────────────┐
│                      UNIT 5                           │
│        RATIONAL EXPONENTS AND RADICALS                │
└─────────────────────────────────────────────────────┘
```

*Unit 2 is a prerequisite for this unit. Answers appear at the end of the unit.

TI-83/83plus users should press [MODE] and verify that **Real** is highlighted in the left column. The TI-83/83plus have complex number capabilities as is indicated by the "a + bi" selection adjacent to **Real**. Complex number operations will be discussed in the unit entitled "TI-83/83plus/85/86 Complex Numbers." For now, the calculator should be set to compute only with real numbers.

Radicals

The inverse operation of raising a number to a power is extracting a root. For example, $4^3 = 64$ and $\sqrt[3]{64} = 4$.

The only type of radical that has been addressed thus far in this text is $\sqrt{}$ or principal square root. We will now examine $\sqrt[3]{}$, $\sqrt[4]{}$, $\sqrt[6]{}$, etc., and the relationship of these radicals to rational exponents.

┌──┐
│ TI-85/86 │ PRESS [2ND] <MATH> TO DISPLAY THE FIRST FIVE SUB-MENUS OF THE TI-85/86'S │
│   │ MATH FEATURE. THE "▸" AFTER **MISC** INDICATES THERE ARE MORE SUB-MENUS. PRESS │
│ THE [MORE] KEY TO DISPLAY THE REMAINING MENU AND THEN PRESS [MORE] ONE MORE │
│ TIME TO RETURN TO THE INITIAL SET OF SUB-MENUS. (SUGGESTION: AFTER COMPLETING │
│ THIS UNIT, EXAMINE THE CONTENTS OF EACH OF THESE SIX SUB-MENUS WHICH ARE │
│ ACCESSED BY PRESSING THE APPROPRIATE F KEY.) THE RADICAL SYMBOL, $\sqrt[x]{}$, THAT WILL │
│ BE USED IN THIS UNIT IS FOUND UNDER THE MISC SUBMENU. PRESS [F5](MISC) FOLLOWED │
│ BY [MORE] TO DISPLAY THE $\sqrt[x]{}$. AT THIS POINT, IT WOULD BE A GOOD IDEA TO RETURN TO │
│ THE CUSTOM MENU AND CUSTOMIZE THE $\sqrt[x]{}$ USING THE CATALOG. │
│ THERE IS NO $\sqrt[3]{}$ AVAILABLE ON THE TI-85/86; CONTINUE THE UNIT ON THE NEXT PAGE. │
└──┘

Begin by looking at the MATH menu. Press [MATH] and the appropriate screen is displayed at the right. To scroll through the entire menu, press the down arrow key. This unit will use the ▸**FRAC** option, as well as the $\sqrt[3]{}$ and $\sqrt[x]{}$ options.

```
┌────────────────────┐   ┌────────────────────┐
│ MATH NUM HYP PRB   │   │ MATH NUM CPX PRB   │
│ 1▪▸Frac            │   │ 1▪▸Frac            │
│ 2:▸Dec             │   │ 2:▸Dec             │
│ 3:³                │   │ 3:³                │
│ 4:³√               │   │ 4:³√(              │
│ 5:ˣ√               │   │ 5:ˣ√               │
│ 6:fMin(            │   │ 6:fMin(            │
│ 7↓fMax(            │   │ 7↓fMax(            │
└────────────────────┘   └────────────────────┘
     TI-82 Screen          TI-83/83plus Screen
```

To find the cube root of -27, $\sqrt[3]{-27}$, press [MATH] [4] (to select the $\sqrt[3]{(}$ option) [(-)] [2] [7] [)] [ENTER]. The result displayed should be -3. Recall, the inverse of extracting a root is raising to a power. Thus -3 is the correct result for $\sqrt[3]{-27}$ because $(-3)^3 = -27$.

To compute seven times the cube root of 5, $7\sqrt[3]{5}$, press [7] [MATH] [4:$\sqrt[3]{(}$] [5]. The result displayed is an approximate answer rounded to nine decimal places: 11.96983163.

To compute roots other than square roots ($\sqrt{}$ is found on the face of the calculator) or cube roots, the $\sqrt[x]{}$ selection on the MATH menu must be used. Designate a value for x by entering the index first and then the $\sqrt[x]{}$ symbol.

To compute the sixth root of 64, $\sqrt[6]{64}$, press [6] [MATH] [5]($\sqrt[x]{}$) [6] [4] [ENTER]. Your screen should match the one at the right.

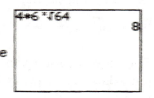

Because the display is $6\sqrt[x]{64}$, care must be taken when entering expressions in which a factor is specified in front of a root. In a problem such as $7\sqrt[3]{5}$, it is unnecessary to "tell" the calculator to multiply 7 times $\sqrt[3]{5}$ because $7\sqrt[3]{5}$ indicates implied multiplication. However in $4\sqrt[6]{64}$ implied multiplication cannot be used because of the calculator notation. Had the expression $4\ 6\sqrt[x]{64}$ been entered, the calculator would compute $\sqrt[46]{64}$. A multiplication symbol, " * ", must be entered after the digit 4 for clarification. Your screen display for $4\sqrt[6]{64}$ should match the one at the right.

NOTE: TI-83/83plus users must be particularly careful when using the $\sqrt[x]{}$. Up to this point, a parenthesis has been automatically entered after the square root or cube root symbol. The $\sqrt[x]{}$ symbol does not include this parenthesis. Any fraction or expression containing more than one character must be enclosed in parentheses when placed under this radical symbol.

Use the calculator to find the value of $\sqrt[4]{\dfrac{16}{625}}$. The correct root is 2/5 (or 0.4). If you got 2/625 (or 0.0032) then you failed to group the fraction with parentheses. Parentheses are necessary for the calculator to extract the fourth root of the quantity 16/625.

$$\boxed{\text{EXERCISE SET}}$$

Directions: Evaluate each radical expression. Record the screen display, being sure that the indicated root is displayed as an integer or fraction - rather than a decimal.

1. $\sqrt[3]{-125}$ 2. $\sqrt[4]{4096}$ 3. $\sqrt[6]{5^6}$

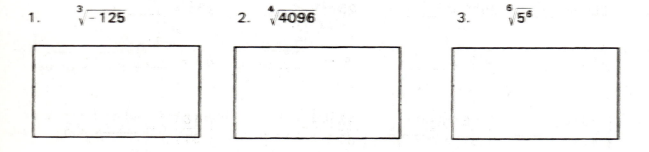

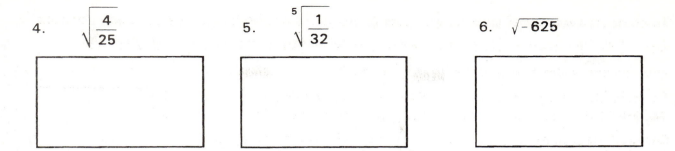

4. $\sqrt{\dfrac{4}{25}}$

5. $\sqrt[5]{\dfrac{1}{32}}$

6. $\sqrt{-625}$

Rational Exponents

We now want to consider the effect of rational (fractional) exponents.

The caret key is used before specifying an exponent (higher than 2), regardless of whether the exponent is an integer or a rational number. However, care must be taken when entering a rational number as an exponent; the order of operations must be considered.

An expression has been entered on the calculator and is displayed on the screen at the right. Two operations are indicated, raising a base to a power and division. Because of the order of operation rules, the base will be raised to the indicated power before the division is performed. This calculator display is not an illustration of $25^{1/2}$ but is

an illustration of $\dfrac{25^1}{2}$. For the calculator to evaluate the expression

$25^{1/2}$, parentheses must be inserted to override the order of operations.

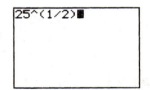

EXERCISE SET

The following problems have been placed in groups of four to more easily compare the effect of differing exponents on the same base. **Be sure all rational exponents are enclosed in parentheses.** Enter final results on the appropriate blank.

7. $25^2 =$ _____ $\quad 25^{1/2} =$ _____ $\quad 25^{-1/2} =$ _____ $\quad 25^{-2} =$ _____

8. $8^3 =$ _____ $\quad 8^{1/3} =$ _____ $\quad 8^{-1/3} =$ _____ $\quad 8^{-3} =$ _____

9. $\left(\dfrac{16}{81}\right)^2 =$ ___ $\quad \left(\dfrac{16}{81}\right)^{\frac{1}{2}} =$ ___ $\quad \left(\dfrac{16}{81}\right)^{-\frac{1}{2}} =$ ___ $\quad \left(\dfrac{16}{81}\right)^{-2} =$ ___

The following exercises allow you to determine the relationship between $\sqrt[n]{a^m}$ and $a^{m/n}$. The problems are grouped in such a way that patterns can be discovered. Record the result on the blank.

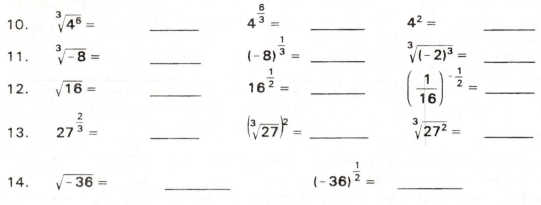

10. $\sqrt[3]{4^6} =$ _____ $4^{\frac{6}{3}} =$ _____ $4^2 =$ _____

11. $\sqrt[3]{-8} =$ _____ $(-8)^{\frac{1}{3}} =$ _____ $\sqrt[3]{(-2)^3} =$ _____

12. $\sqrt{16} =$ _____ $16^{\frac{1}{2}} =$ _____ $\left(\frac{1}{16}\right)^{-\frac{1}{2}} =$ _____

13. $27^{\frac{2}{3}} =$ _____ $\left(\sqrt[3]{27}\right)^2 =$ _____ $\sqrt[3]{27^2} =$ _____

14. $\sqrt{-36} =$ _____ $(-36)^{\frac{1}{2}} =$ _____

✎15. Explain the relationship between $\sqrt[n]{a^m}$ and $a^{m/n}$.

✎16. In #14 the TI-82/83/83plus calculator displayed a DOMAIN ERROR. Explain what is wrong with the problem.

Solutions: 1. -5   2. 8   3. 5   4. 2/5   5. ½

6. Did you get an error message? The DOMAIN error message is displayed because the TI-82 only computes values of real numbers and TI-83/83plus users were instructed to set the calculator to compute only with real numbers. In the real number system, $\sqrt{-625}$ does not exist. There is no real number that when raised to the second power will equal -625.
The TI-85/86 will display (0,25) instead of an ERROR message as the TI-82/83 does. This is because the TI-85/86 can perform complex number operations. The $\sqrt{-625}$ simplifies to the complex number $0 + 25i$. The calculator notation for $0 + 25i$ is (0,25). The use of the TI-85/86 in the simplification of complex number expressions is discussed in detail in the unit entitled "TI-83/85/86 Complex Numbers." Until the instructor and/or textbook addresses complex numbers, results displayed in the form (a, b) should be recorded as *no real number*.

7. 625, 5, 1/5, 1/625   **8.** 512, 2, 1/2, 1/512   **9.** 256/6561, 4/9, 9/4, 6561/256
10. 16, 16, 16   **11.** -2, -2, -2 **12.** 4, 4, 4   **13.** 9, 9, 9
14. not a real number, not a real number

| TI-85/86 | TI-85/86 USERS WILL HAVE (0,6) FOR BOTH RESULTS IN 14. REMEMBER, THIS IS THE CALCULATOR'S NOTATION FOR COMPLEX NUMBERS WHICH WILL BE ADDRESSED IN A LATER UNIT. |

15. In the fractional exponent $\frac{m}{n}$, the n designates the index of the radical and the m is the exponent applied to the radical.

16. The square root function is not defined for negative radicands.

38

*Unit 3 is a prerequisite for this unit. Answers appear at the end of the unit.

NOTE: The TI-82 does not address complex numbers.

Complex numbers are numbers that can be expressed in the form a + bi where **a** represents the real part and **b** represents the imaginary part.

TI-83/83plus   The TI-83/83plus has the value "*i*" on the calculator face, located above the decimal point. This allows the easy entry and arithmetic manipulation of complex number expressions.

1. Complex numbers can be represented by variables, that is to say they can be stored as variables. Let X = 4 - 2i and evaluate $3X^2 + 2X - 5$. Using the **STO►** key, your display should look like the one at the right. The interpretation of the display is that replacing X by the complex number 4 - 2i in the expression $3X^2 + 2X - 5$, yields a value of 39 - 52i.

```
4-2i→X:3X²+2X-5
            39-52i
```

2. Access the complex number menu by pressing **[MATH] [►] [►]** to highlight **CPX**. This menu indicates that for any complex number, a + bi, the calculator will identify the conjugate, determine the real part, determine the imaginary part, and determine the absolute value. The **►FRAC** option can be used with any of the operations.

```
conj(1/2-3i)
          .5+3i
real(1/2-3i)
          .5
imag(1/2-3i)
          -3
```

```
abs(1/2-3i)
      3.041381265
```

3. Arithmetic operations with complex numbers can be performed easily with this calculator. For example, to perform the division $\frac{2 + 3i}{3 + 2i}$ by hand, the numerator and denominator of the fraction must be multiplied by the conjugate of the denominator in order to simplify the expression. Be sure to use parentheses around both the numerator and denominator when entering the fraction in the calculator and use the **►FRAC** option to express the quotient in fraction form. The displayed result should correspond to the one pictured.

```
(2+3i)/(3+2i)►Fr
ac
       12/13+5/13i
```

TI-85/86   COMPLEX NUMBERS ARE DISPLAYED AS **(REAL, IMAGINARY)** ON THE TI-85/86. THUS 2 - 3i WOULD BE DISPLAYED AS (2,-3) SINCE 2 IS THE REAL PART AND -3 IS THE IMAGINARY PART. **NOTE:** (REAL, IMAGINARY) IS RECTANGULAR FORMAT, NOT POLAR IN THE UNIT ENTITLED "THE STORE AND TABLE FEATURES", PG.28, THE USE OF LOWERCASE LETTERS AS SYSTEM RESERVED VARIABLES IS DISCUSSED. THE LOWERCASE i CAN BE DESIGNATED AS A USER DEFINED VARIABLE IN THE TI-85/86 BY USING THE CONSTANT EDITOR. USE OF THE CONSTANT EDITOR TO DEFINE i AS (0,1), REPRESENTING 0 + i, PREVENTS YOU FROM STORING AN ALTERNATE VALUE IN i AT A LATER DATE.

PRESS [2ND] <CONS> [F2](EDIT) AND ENTER A LOWERCASE **i** AFTER **Name** = BY PRESSING [2ND] <ALPHA> <I>. CURSOR DOWN TO **Value** = AND ENTER **(0,1)** TO DEFINE THE VALUE AS THE COMPLEX NUMBER 0 + i. COMPLEX NUMBERS MAY NOW BE ENTERED ON THE HOME SCREEN IN A + Bi FORMAT INSTEAD OF (A,B) FORMAT IF DESIRED. HOWEVER, ANSWERS WILL BE RETURNED BY THE CALCULATOR IN THE (A,B) FORMAT.

1. COMPLEX NUMBERS CAN BE REPRESENTED BY VARIABLES, THAT IS TO SAY THEY CAN BE STORED AS VARIABLES. LET X = 4 - 2i AND EVALUATE $3x^2 + 2x - 5$. USING THE **STO►** KEY, THE DISPLAY SHOULD LOOK LIKE THE ONE AT THE RIGHT. INTERPRETING THE DISPLAY, WE SEE THAT WHEN X = 4 - 2i, $3x^2 + 2x - 5$ HAS A VALUE OF 39 - 52i.

2. PRESS [2ND] <CPLX> TO ACCESS THE COMPLEX NUMBER MENU. FOR ANY COMPLEX NUMBER, A + Bi, THE CALCULATOR WILL IDENTIFY THE CONJUGATE (F1), DETERMINE THE REAL PART (F2), DETERMINE THE IMAGINARY PART (F3), AND DETERMINE THE ABSOLUTE VALUE (F4). THE REMAINING OPTIONS ARE USED WITH POLAR FORMAT. THE DISPLAYS AT THE RIGHT INDICATE THE APPLICATIONS OF EACH OF THESE OPTIONS WHEN APPLIED TO ½ - 3i. THE **►FRAC** OPTION CAN BE USED WITH ANY OF THESE OPERATIONS.

3. ARITHMETIC OPERATIONS WITH COMPLEX NUMBERS CAN BE PERFORMED EASILY WITH THE TI-85/86. FOR EXAMPLE, TO PERFORM THE DIVISION $\frac{2 + 3i}{3 + 2i}$ BY HAND, YOU WOULD NEED TO MULTIPLY THE NUMERATOR AND DENOMINATOR BY THE CONJUGATE OF THE DENOMINATOR IN ORDER TO SIMPLIFY THE EXPRESSION. THE CALCULATOR ALLOWS YOU TO ENTER THE PROBLEM AS DISPLAYED ON THE SCREEN AT THE RIGHT. BE SURE TO USE THE **►FRAC** OPTION TO EXPRESS THE QUOTIENT IN FRACTION FORM. THE DISPLAYED RESULT IS $\frac{12}{13} + \frac{5}{13}i$ IN STANDARD FORM.

EXERCISE SET

Directions: Use the complex number format and menu on the calculator to perform the complex number operations below. Copy your screen display and record all answers in standard a + bi form.

1.   (3 + 2i) + 4(-2 + 5i)   1._____

2.   (3 + 2i)(-2 + 5i)   2._____

3. $\dfrac{5 + i}{2 - 3i}$ 　　　　　　　　　　　　　3._____

4. $(2 - 3i)^3$ 　　　　　　　　　　　　　4._____

5. i^{23} 　　　　　　　　　　　　　5._____

6. i^{114} 　　　　　　　　　　　　　6._____

7. Use the **STO▶** feature to show that $\pm 5i\sqrt{2}$ is a solution to $x^2 + 50 = 0$.

8. $\left| \dfrac{4}{5} + \dfrac{2}{3}i \right|$ 　　　　　　　　　　　　　8._____

Can the result be converted to a fraction?_____

✍ Compute this same problem by hand and explain why.

9.   Simplify the following on the calculator:

    a. $4 + \sqrt{-4}$ 9a._____

    b. $\dfrac{2}{5} + \sqrt{-\dfrac{4}{9}}$ (Be sure the answer is expressed in fractional form.)   9b._____

10.   Evaluate $\sqrt{-2}$ on the calculator. Express the result as a fraction, or explain why the calculator will not convert it to a fraction.

Solutions: **1.** -5 + 22i   **2.** -16 + 11i   **3.** $\dfrac{7}{13}$ + $\dfrac{7}{13}i$ **4.** -46 - 9i

5. TI-83: 3E -13 -i which is equivalent to 0 - i
   TI-85: (3E -13, -1) which is equivalent to (0,-1) which is 0 - i in standard form

6. TI-83: -1 + 1.4E -12 which is equivalent to -1 + 0i or -1
   TI-85: (-1, 1.4E -12) which is equivalent to (-1,0) which is -1 + 0i in standard form or -1.

7.
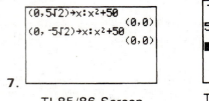
TI-85/86 Screen

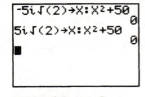
TI-83/83plus Screen

8. 1.04136662345: This number cannot be converted to a fraction because it is an irrational number.

9a. 4 + 2i   **9b.** $\dfrac{2}{5}$ + $\dfrac{2}{3}i$ **10.** The number cannot be converted to a fraction because it is an irrational number.

43

UNIT 7
GRAPHICAL SOLUTIONS: LINEAR EQUATIONS

*Unit 3 is a prerequisite for this unit. However, depending on the organization of your textbook you may choose to work units 16 and 17 prior to this unit. Answers appear at the end of the unit.

This unit explores the use of the INTERSECT feature for solving equations. Recall, an equation is a statement that two algebraic expressions are equal. Solving an equation means finding a replacement value for X that will produce the same value for both expressions. This unit will consider first degree conditional equations in which there will be one correct solution, identities for which there are infinitely many solutions and contradictions for which there are no solutions.

Solve 3 - 2X = 2X + 7, algebraically: Check the solution by substitution:

The solution of the equation is _____.
When this value for X is substituted into the expression 3 - 2X, that expression evaluates to _____. When this value for X is substituted into the expression 2X + 7, that expression evaluates to _____. Thus, the left and right sides of the equation have the same value of 5 when X is equal to -1.

TI-85/86	TI-85/86 USERS GO TO THE GUIDELINES WHICH FOLLOW THIS UNIT (PG.53).

We will now look at the graphical representation of each side of the equation AND interpret the graph screen to determine the solution of the equation. Press [Y =] and [CLEAR] to delete any expressions. The left side of the equation, 3 - 2X, will be designated as Y1. Enter 3 - 2X after "Y1 =" (remember to use the [X,T,θ,n] key). The right side of the equation, 2X + 7, will be designated as Y2. Enter 2X + 7 after "Y2 =". We want to determine graphically where Y1 = Y2.

Press [WINDOW] and cursor down to each line to enter the values displayed on the adjacent screen. The resolution (Xres) at the bottom of the screen should be equal to 1. If not, cursor down to change it.

```
WINDOW
Xmin=-9.4
Xmax=9.4
Xscl=1
Ymin=-6.2
Ymax=6.2
Yscl=1
Xres=1
```

TI-82   When [WINDOW] is pressed on the TI-82, the blinking cursor appears on WINDOW. Cursor down to enter the values displayed.

```
        FORMAT
Xmin=-9.4
Xmax=9.4
Xscl=1
Ymin=-6.2
Ymax=6.2
Yscl=1
```

These values designate the size of the viewing window when graphing Y1 and Y2. These WINDOW values were chosen to give "friendly" numbers when the TRACE feature is activated later in this unit.

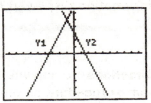

Press [**GRAPH**]. The graphical solution to the equation is the intersection of Y1 and Y2. At the intersection point, Y1 is equal to Y2. Circle the intersection point.

To interpret this graph, press [**TRACE**] and move along the graph to the intersection point. The left and right arrow keys will move the cursor along the path of the graph. The equation in the upper left corner indicates which graph is being traced, either Y1 or Y2.

TI-82	When [**TRACE**] is pressed, the TI-82 displays (in the upper right corner) the small "1" or "2" to indicate which graph is being traced.

To ensure the cursor is on the <u>exact</u> point of intersection, record the values indicated at the bottom of the screen:_____. Press the up (or down) arrow key once to move to the other graph. The equation (or number for TI-82 users) at the top of the screen will change, even though the cursor will not appear to move. Compare the numbers that are now displayed at the bottom of the screen to those recorded above. If they are <u>both</u> the same, then the cursor is on the exact point of intersection. If not, adjust the TRACE cursor and test again.

At the intersection point, X = - 1 and Y = 5. Remember, the equation 3 - 2X = 2X + 7 evaluates to a true arithmetic statement when X has a value of - 1. Because we entered 3 - 2X at the Y1 = prompt and 2X + 7 at the Y2 = prompt the y-value of 5 tells us that each side of the original equation evaluates to 5 when X is replaced by - 1. This should agree with the algebraic solution computed at the beginning of the unit.

Intersect Option

Because tracing to a point of intersection is dependent on appropriate WINDOW values, the INTERSECT option will be used exclusively from this point on. It is not dependent on the window values entered. The only requirement is that the point of intersection be visible on the screen. Check your WINDOW values to ensure they are the same as those displayed on the screen at the right. Press [**WINDOW**] and use

```
WINDOW
 Xmin=-10
 Xmax=10
 Xscl=1
 Ymin=-10
 Ymax=10
 Yscl=1
 Xres=1
```

the down arrow key to move down the screen and change the values, if necessary.

TI-85/86	TI-85/86 USERS SHOULD RETURN TO THE GUIDELINES WHICH FOLLOW THIS UNIT (PG.53).

To access the INTERSECT option, press [**2nd**] <**CALC**>. (**CALC** is located above TRACE.) Press [**5:intersect**] to select INTERSECT. Move the cursor along the first curve to the approximate point of intersection and press [**ENTER**]. At the *second curve* prompt press [**ENTER**] again because the cursor will still be close to the point of intersection. At the *guess* prompt, press [**ENTER**]. This is instructing

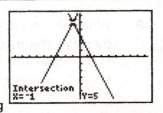

the calculator that our *guess* is the approximate point of intersection that was designated at the *first curve* prompt. The solution of the equation is - 1, the displayed x-value.

Now check this solution. Refer back to the unit entitled "The STOre and TABLE Features," if you do not remember how to do this. Compare the results of your check work to the information displayed on the INTERSECT screen.

EXERCISE SET

Directions: Using the INTERSECT option, solve these equations graphically. Follow the same procedure as outlined. Sketch the INTERSECT screen that yields the solution and use ►Frac to convert all decimal answers to fractions.

1. $5 = 2 - 7X$

   $X =$ _____

Note: To convert an x-value on the INTERSECT screen to a fraction you must return to the Home Screen. Press **[2nd]** <**QUIT**>. The value of **X** computed by the INTERSECT feature is stored in X. To retrieve this value, enter X (press **[X,T,θ,n]**) and then convert its value to a fraction, if necessary, by pressing **[MATH] [1:►Frac] [ENTER]**.

   Converted to a fraction, $X =$ _____

2. $(-4/3)X = -2$

   $X =$ _____
   Converted to a fraction, $X =$ _____

*(In your textbook, this problem would look like this: $-\frac{4}{3}x = -2$)

3. $\dfrac{-4X}{3} + 6 = -1$

   $X =$ _____
   Converted to a fraction, $X =$ _____

4. $\dfrac{2X - 1.2}{0.6} = \dfrac{4X + 3}{-1.2}$

   $X =$ _____
   Converted to a fraction, $X =$ _____

Did you get -3/40 (i.e. -.075)? If you did not, check the way the equation was entered. Parentheses will need to be inserted in the appropriate places.

✎5.   When the equation $7x + 8 = -9 - 3x$ is solved graphically, the bottom of the screen displays the message "INTERSECTION x = -1.7   y = -3.9". Explain the meaning of the X and Y values within the context of the equation that was being solved.

NOTE: The equations in this unit were specifically written to conform to the WINDOW values displayed on the screen at the right. This set of window values is designated as the standard viewing window. In order to maintain a consistent point of reference, all graphical solutions to equations will begin by displaying the graph in the standard viewing window. This can be easily set by pressing [ZOOM] [6:ZStandard].

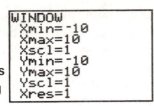

(TI-85/86 users press [GRAPH] [F3](ZOOM) [F4](ZSTD).) For further discussion of calculator viewing windows, you may want to do the units entitled "How Does the Calculator Actually Graph?" followed by "Preparing to Graph: Calculator Viewing Windows".

Recall that the axes seen displayed on the graph screen are simply two number lines placed perpendicularly at their origins. The horizontal number line is the X-axis and the vertical number line is the Y-axis. They are oriented so that right and up are positive directions. The term Xmax on the WINDOW screen refers to the maximum distance from the origin (0) to the right side of the viewing window, whereas Xmin refers to the distance from the origin to the left side of the window. Ymax is the distance from the origin to the top of the viewing window, whereas Ymin is the distance from the origin to the bottom of the screen.
It is best that the point(s) of intersection be visible in the viewing window if using the INTERSECT option. When not visible, one or both of the axes can be stretched.

Example 1: Solve the equation 3(X - 4) = 6 + X graphically, using the INTERSECT feature.

Solution: After entering the left and right sides of the equation at the appropriate prompts and pressing **[GRAPH]** , the displayed graph appears. The viewing WINDOW is not sufficiently large to allow the display of the point of intersection.
Which part of which axis must be stretched? by how much?

The Ymax must be made larger, and it is suggested that the student simply make an educated guess. After accessing the WINDOW menu, change Ymax to 20. The point of intersection should be clearly visible as illustrated by the second screen. Use of the INTERSECT option displays the correct solution of 9. ◆

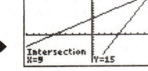

EXERCISE SET CONT'D

Directions: Solve each equation using the INTERSECT feature of the calculator. Sketch the graph displayed, including the axes. Record window values in the spaces provided. It is suggested that the standard viewing window, given above, be used when first graphing, and then modifications can be made as appropriate.

6.   3(X + 2) - 9 = 15

       X = _____

       WINDOW values:

       XMin: _____   XMax: _____   YMin: _____   YMax: _____

7. $3(X - 4) + 7 = 2(X + 4)$

   X = _____

   WINDOW values:

   XMin: _____   XMax: _____   YMin: _____   YMax: _____

8. $\dfrac{X}{2} + \dfrac{X}{2} = 20$

   X = _____

   WINDOW values:

   XMin: _____   XMax: _____   YMin: _____   YMax: _____

NOTE: When using the graph screen to solve equations/inequalities, you should be aware that the displayed coordinate values approximate the actual mathematical coordinates. The accuracy of the displayed values is determined by the height and width of the pixel space being displayed. The space height/width formulas are discussed in detail in the unit entitled "Preparing to Graph: Viewing Windows".

Special Cases

Example 2: Solve $4(X - 1) = 4X - 4$ graphically.

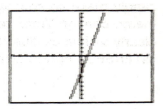

Solution: Press [Y =] and enter $4(X - 1)$ after Y1 = and $4X - 4$ after Y2 =. Press [TRACE]. ONLY ONE graph is displayed!!! Trace along this line and observe the equation in the upper left corner of the screen. Which graph are you tracing on? Now use the up (or down) arrow key (press only one) and move the TRACE cursor to the other graph. Again check the equation displayed in the upper left corner or the number in the upper right corner, depending on the calculator. Which graph are you on now? Both graphs are the same! When both graphs are the same, then both sides of the equation must be equivalent expressions.

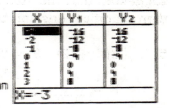

Equivalent expressions produce identical values for all replacement values of the variable. Look at the table of values at the right that displays both the left and right side of the equation. Clearly, both sides of the equation are equivalent for each value of x. Simplifying each side of the equation algebraically justifies that the equation is an identity. Identities are true for all values of X that are acceptable replacement values for the variable in the equation. Thus the solution to this equation is the set of all real numbers.

◆

49

Example 3: Solve 2X - 5 = 2(X + 1) graphically.

Solution: Press [Y =] enter 2X - 5 after Y1 = and 2(X + 1) after Y2 = . Press [**TRACE**]. Observe that the two lines appear to be parallel. Parallel lines never intersect and hence there is no solution. To indicate that this equation has no solution, the symbol for the empty set is written, { } or ∅. Upon simplifying both sides of the original equation it is apparent algebraically there is no solution. This equation is called a contradiction. ◆

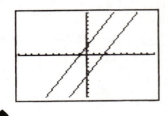

EXERCISE SET CONT'D

9. Graphical representations of the three types of equations are displayed below. In the blank provided, identify each type of equation.

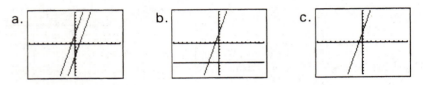

a. _____   b. _____   c. _____

10. The table below represents one of the three types of equations. Determine the type of equation *and* the solution to the equation.

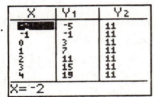

    type:_____   solution:_____

✍11. IN YOUR OWN WORDS: Write a summary of what you learned in this unit. You should address the following:
    a. the manner in which you enter equations on the calculator,
    b. the use of the INTERSECT option,
    c. discuss stretching axes,
    d. graphical representation of the three types of linear equations (conditionals, identities, and contradictions)
    e. illustrate the use of the STOre and TABLE features of the calculator as applicable to checking solutions of equations.

Solutions to Exercise Sets: **1**. -.4285714, $-\dfrac{3}{7}$ **2.** 1.5, $\dfrac{3}{2}$ **3.** 5.25, $\dfrac{21}{4}$ **4.** -.075, $-\dfrac{3}{40}$

5. The X-value is the solution to the equation. The Y-value is the quantity each side of the equation will evaluate to when the variable is replaced with the solution value.

6. 6; Xmin = -10, Xmax = 10, Ymin = -10, Ymax = at least 16

7. 13; Xmin = -10, Xmax = at least 14, Ymin = -10, Ymax = at least 35

8. 20; Xmin = -10, Xmax = at least 21, Ymin = -10, Ymax = at least 21

9a. contradiction   **9b.** conditional   **9c.** identity

10. Type: conditional; Solution: 2

To graphically solve this same equation (3 - 2X = 2X + 7), the graphical representation of the algebraic expression of each side of the equation will be examined.

To graph, press **[GRAPH]**. This will bring up a menu at the bottom of the screen. These options correspond to the five buttons at the top of the TI-82/83/83plus, with RANGE corresponding to WINDOW if you are using a TI-85, specified as WIND on the TI-86.

To graphically solve 3 - 2X = 2X + 7, press **[F1](y(x) =)**. Enter 3 - 2X (the left side of the equation) at the y1 = prompt that is displayed (enter the variable x by pressing the **[x-VAR]** key). Pressing **[ENTER]** after typing in the expression will automatically display y2 = . Enter 2X + 7 at this prompt.

Press **[2nd]** **<M2>(RANGE)**, **WIND** on the TI-86, and cursor down to enter the following values for a "friendly" viewing window. TI-86 users be sure that xRes is set equal to 1.

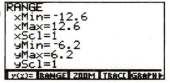

Press **[F5](GRAPH)**.

☞ RETURN TO PG.46 IN THE CORE UNIT AND CONTINUE READING.

To access the **ISECT** option (INTERSECT) begin by displaying the graph: press **[2nd]** **<M5>(GRAPH)**. Your display should match the one pictured. The solution of the equation 3 - 2x = 2x + 7 is the x-value of the point on the graph where the two lines intersect. We will use the **ISECT** option (intersect) of the calculator to find this x-value.

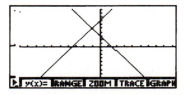

Press **[MORE]** to see more options on the menu. The left option is **MATH**. Press **[F1](MATH)** to select the **MATH** menu, then **[MORE]** and select **(ISECT)** for intersection. (TI-86 users should return page 46 in the core unit and follow the prompts.) With the cursor near the desired point of intersection, press **[ENTER]**. The cursor actually moves from one graph to the other with this first press of **ENTER**. If necessary, again move the cursor close to the point of intersection and press **[ENTER]** for the second time. This press of **ENTER** activates the **ISECT** computation. Your screen should correspond to the one at the right.

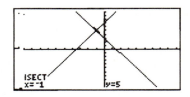

   Note: If there is more than one point of intersection, the **ISECT** function must be completed for each point of intersection.

(Pressing **[EXIT]** or **[CLEAR]** removes the menus displayed at the bottom of the graph.)

☞ RETURN TO PG.47 AND CONTINUE READING.

UNIT 8
GRAPHICAL SOLUTIONS: ABSOLUTE VALUE EQUATIONS

*Unit 7 is a prerequisite for this unit. Answers appear at the end of this unit.

In the prerequisite unit, linear equations were solved graphically using the INTERSECT feature of the calculator. This same approach will be used to solve equations containing absolute value. Remember, absolute value is located by pressing **[MATH] [▶]** for the **NUM** menu or by pressing **[2nd] <CATALOG>**. (TI85/86 users should have this feature customized).

 Remember, absolute value is located above the x^{-1} key and is accessed by **[2nd] <x⁻¹>**.

Consider the equation $|X + 3| = 6$. To solve graphically, enter the left side of the equation as **abs(X + 3)** at the Y1 = prompt and the right side of the equation at Y2 =. (Remember, $|X + 3|$ is read as "the absolute value of the quantity X plus three".) Press **[ZOOM]** **[6:ZStandard]** (TI-85/86 users press **[GRAPH] [F3](ZOOM) [F4](ZSTD)**) to automatically enter the values on the WINDOW screen at the right .

```
WINDOW
 Xmin=-10
 Xmax=10
 Xscl=1
 Ymin=-10
 Ymax=10
 Yscl=1
 Xres=1
```

The keystrokes display the graph; confirm these WINDOW values by pressing **[WINDOW]**.

Press **[GRAPH]** to display the graphical representation of the absolute value equation. **See** the display at the right.

    a.   Circle the two points of intersection that are displayed.

    b.   Use the calculator's INTERSECT option to find the X values at the points of intersection. You will need to use the INTERSECT option <u>twice</u> since there are two points of intersection. Copy the screen display to show where you found <u>one</u> of the two solutions, but record both of the solutions here: X = _____ or X = _____

NOTE CONCERNING THE USE OF THE INTERSECT OPTION: In the future there may be other equations that have more than one solution. There will be an intersection point for each of these solutions that is a real number. The INTERSECT option will have to be completed for each point of intersection. To justify your work you will only be required to sketch one of the INTERSECT screens that you used and merely record the solutions derived from the other screens.

Confirm the solutions of 3 and -9 in the following ways:
    a. analytically through substitution   b. via the TABLE

EXERCISE SET

Directions: Graphically solve each of the equations below. Sketch the screen. Circle the points of intersection. Use the INTERSECT option to find the intersections. REMEMBER: Because there are two points of intersection, the process will have to be done twice. Record both of the solutions on the blanks provided.

1. $|2X - 1| = 5$

    X = _____ or X = _____

2. $\left|\dfrac{1}{2}X + 1\right| = 3$

    X = _____ or X = _____

3. $\left|\dfrac{4 - X}{2}\right| = \dfrac{8}{5}$

    X = _____ or X = _____

4. $|X + 1| = |2X - 3|$

    X = _____ or X = _____

5. $|2 - X| = |3X + 4|$

    X = _____ or X = _____

6. $|4X + 5| = -2.$

    Do the graphs intersect? _____

    What is the solution? _____

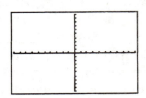

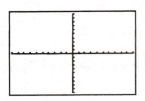

56

✎ 7. You should be able to determine the solution to $|4X + 5| = -2$ by merely <u>looking</u> at the problem. What clue lets you know that there is no solution?

Directions: Graphically solve each of the equations below. Sketch the screen (axes will need to be adjusted and WINDOW values recorded in the spaces provided). Circle the points of intersection. Use the INTERSECT option to find the intersections. REMEMBER: Because there are two points of intersection, the process will have to be done twice. Record both of the solutions on the blanks provided.

8. $|2(X - 5) - 9| = 5$

Xmin = _____ , Xmax = _____, Ymin = _____, Ymax = ____

X = _____ or X = _____

9. $|2X + 7| = 11$

Xmin = _____ , Xmax = _____, Ymin = _____, Ymax = ____

X = _____ or X = _____

10. $|4X - 3| = |2x + 5|$

Xmin = _____ , Xmax = _____, Ymin = _____, Ymax = ____

X = _____ or X = _____

11. $|3X - 1| = |7 + 4X|$

Xmin = _____ , Xmax = _____, Ymin = _____, Ymax = ____

X = _____ or X = _____

12. $|X - 3| = -|X + 20|$

Xmin = _____ , Xmax = _____, Ymin = _____, Ymax = _____

Do the graphs intersect?_____

What is the solution?_____

Hint: Try zooming out by pressing **[ZOOM] [3:Zoom Out] [ENTER]**. TI-85/86 users press **[GRAPH] [F3] (ZOOM) [F3] (ZOUT) [ENTER]**.

✍13. IN YOUR OWN WORDS: Briefly describe the process of solving absolute value equations using the INTERSECT option. Your discussion should center on when there are two solutions, one solution, and no solution to the equation.

Solutions:

Exercise Set: 1. X = -2 or X = 3, **2**. X = -8 or X = 4, **3**. X = 0.8 or X = 7.2, **4**. X = 2/3 or X = 4

5. X = -3 or X = -0.5 **6**. No, null set **7**. The solution is the empty set because the two graphs do not

intersect. **8**. X = 7 or X = 12, Xmax should be larger than 12

9. X = -9 or X = 2, Ymax at least 12 **10**. X = -1/3 or X = 4, Ymax at least 14

11. X = -8 or X = -6/7, Ymax at least 26 **12**.. Xmin = -30, no, φ,

58

UNIT 9
GRAPHICAL SOLUTIONS: QUADRATIC
AND HIGHER DEGREE EQUATIONS

*Unit 7 is a prerequisite for this unit. Answers appear at the end of this unit.

FACTORABLE EQUATIONS

Polynomial equations can be solved by the INTERSECT method that was used in previous units. Both sides of the equation can be graphed and the solution(s) determined from the point(s) of intersection. There is, however, another graphic approach to solving equations that will be considered in this unit. This method is the ROOT or ZERO method and can be used to find the REAL roots/zeroes of all the equations you have learned to solve graphically thus far and for any equation encountered in the future. Calculator techniques introduced early in the unit focus on quadratic equations. Later exercises expand the techniques to higher degree polynomial equations. Thus, the unit can easily be divided into two parts if your text addresses quadratic equations in a section separate from higher order equations.

INTERSECT METHOD

First, a review of the INTERSECT method: This method is best applied to problems in which the equation is not set equal to zero. Remember, it is critical when using this method that all point(s) of intersection are visible. These points of intersection are the solutions/roots/zeroes of the equation.

Example 1: Graphically solve $2(X^2 - 3) - 2X = 3(X + 2)$.

Solution: To solve $2(X^2 - 3) - 2X = 3(X + 2)$ you will need to press [Y =] and enter $2(X^2 -3) - 2X$ after Y1 = and $3(X + 2)$ after Y2 = . Display the graphs in the standard viewing window. One point of intersection is obvious, but the other point is out of the viewing window. (See the screen at the right.) Since the graph appears to have another intersection point "above" the viewing window, press [**WINDOW**] and adjust the size of the viewing window. Change the Ymax to 20 (an educated guess) instead of 10. Press [**GRAPH**] and your screen should look like the one displayed at the right.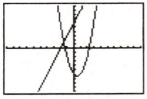

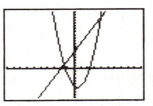

Because the INTERSECT option works independently of the viewing window, the WINDOW may be adjusted as necessary for the given equation. As long as all points of intersection are visible on the display, the calculator will compute the value for each point of intersection. Two points of intersection are displayed. The INTERSECT option discussed in the next paragraph will need to be performed for each point of intersection.

TI-85/86	IF MORE THAN ONE LINE OF MENUS ARE DISPLAYED, PRESS [**EXIT**] TO ENSURE ONLY ONE LINE OF MENU OPTIONS IS DISPLAYED. PRESS [**MORE**] [**MATH**] [**MORE**] AND THE APPROPRIATE F KEY TO ACCESS THE **ISECT** OPTION.

To access the INTERSECT option, press [2nd] <CALC>. (CALC is located above TRACE.) Press [5:intersect] to select INTERSECT. Move the cursor along the first curve to the approximate location of one point of intersection and press [ENTER]. At the "second curve" prompt, press [ENTER] again because the cursor will still be close to the point of intersection. At the "guess" prompt, press [ENTER]. This is instructing the calculator that our "guess" is the approximate point of intersection that was designated at the "first curve" prompt. Your screen should correspond to the one adjacent when you compute the far right point of intersection. Repeat the process for the remaining point of intersection to determine that the two solutions are 4 and -1.5.

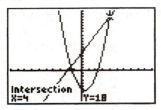

Example 2: Use the INTERSECT option to solve
2(X + 2)(X - 2) + 4 = (X + 4)(X - 1) - 2X graphically.

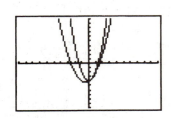

Solution: After the expressions have been entered at the Y1 = and Y2 = prompts, view them in the standard viewing window. Your screen should look like the screen at the right.

To help us determine the points of intersection, we will use the **ZOOM IN** option to get a closer look at the section where the two graphs appear to intersect. Press [ZOOM] [2:Zoom In].

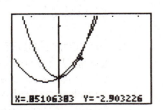

 TI-85/86 USERS PRESS [F3](ZOOM) [F2](ZIN).

Use the down arrow key to cursor down to the third tic mark on the vertical axis (where Y = -3) and then the right arrow key to locate the cursor over the graph. (NOTE: The cursor is moved to a point in the neighborhood of the intersection of the two graphs so the calculator will ZOOM IN on that specific region.) Now press [ENTER]. Your graph display should look very much like the one above. Two points of intersection are now visible.

Use the INTERSECT option to compute both points of intersection and thus determine the two solutions of the equation (i.e. roots). The roots are 0 and 1.

ROOT/ZERO METHOD

TI-82 This option is called "root" and is the second option under the **CALC** menu (i.e. the same relative position as the "zero" option on the TI-83/83plus).

There is an alternate method for graphically solving equations with real roots. This method uses the ROOT (TI-82/85/86) or ZERO (TI-83/83plus) option. If an equation is set equal to zero, when the non-zero side of the equation is graphed, the real roots or zeroes are the X-values at the point(s) where the graph crosses the horizontal axis (X-axis). This method is _perfect_ for equations of any type that are already set equal to zero.

If an equation is not set equal to zero, then you should try the INTERSECT method first. WHY? Because often when you begin to algebraically manipulate an equation you make careless errors. By merely entering the existing equation into the calculator you run less risk of error. If the points of intersection are not clearly visible then you can set the equation equal to zero and use the ROOT option.

Example 3: Solve the previous equation by the ROOT/ZERO method.

Solution: Begin by setting $2(x + 2)(x - 2) + 4 = (x + 4)(x - 1) - 2x$ equal to zero. We will perform the **least** amount of algebra possible to accomplish this:

$$2(x + 2)(x-2) + 4 = (x + 4)(x-1) - 2x$$

$$2(x + 2)(x-2) + 4 \underline{+ \ 2x} = (x + 4)(x-1) - 2x \underline{+ \ 2x} \qquad \text{a. add 2x to both sides}$$

$$2(x + 2)(x-2) + 4 + 2x = (x + 4)(x-1)$$

$$2(x + 2)(x-2) + 4 + 2x \underline{- (x + 4)(x-1)} = (x + 4)(x-1) \underline{- (x + 4)(x-1)} \qquad \text{b. subtract (x + 4)(x-1) from both sides}$$

$$2(x + 2)(x-2) + 4 \underline{\ + \ 2x - (x + 4)(x-1)} = 0 \qquad \text{c. compare the \underline{underlined} part to the original equation}$$

It is this algebraic process of setting the expression equal to zero that encourages us to use the INTERSECT option if at all possible when the equation is not already set equal to 0!

OPTIONAL NOTE: If $2(X + 2)(X-2)$ is entered at Y1 = and $(X + 4)(X-1) - 2x$ is entered at Y2 =, then Y1-Y2 represents the non-zero side of the above equation. Entering Y1-Y2 after Y3 = accomplishes the same result as the above algebraic manipulation. The advantage is that all possibility of error by hand computation is eliminated. If this approach is used, be sure to "turn off" the graphs of Y1 and Y2 by placing the cursor over the equal sign and pressing [ENTER]. (TI-85/86 users press [F5](SELCT) from the y(x) = menu.) This ensures that only the graph of Y3 is displayed.

Enter the NON-zero side of the equation after Y1 = and display in the standard viewing window. The graph is displayed at the right.

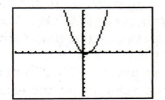

The curve appears to "dip" slightly below the horizontal axis. To get a better view of this section of the curve, use the calculator's **ZBox** option to box in this section. This will alter the WINDOW values.

TI-85/86	ZBOX IS DESIGNATED AS "BOX" ON THE ZOOM MENU. PRESS [ZOOM] [F1](BOX) TO ACCESS THE BOX OPTION AND THEN FOLLOW THE DIRECTIONS IN THE FOLLOWING PARAGRAPH.

Press **[ZOOM] [1:ZBox]**. To box in the area around the root(s), use the arrow keys to move the blinking cursor to the upper left hand corner of the area to be boxed in. Press **[ENTER]**. Use the right arrow key to establish the width of the box followed by the down arrow to establish the height of the box. Your screen should be similar to the one displayed at the right.

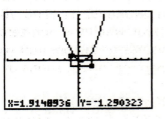

61

The calculator will now enlarge the boxed in area. Press **[ENTER]** to activate. Your screen should be similar to the one at the right.

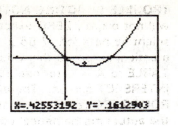

NOTE: Rather than the **ZBox** option, the WINDOW values could have been adjusted using the methods described in Example 1, or the ZOOM IN option could have been used. You may want to explore both the ZBOX and the ZOOM IN options of the ZOOM menu.

Solutions will be found by accessing the ROOT/ZERO option on the calculator.

 BEFORE ACCESSING THE ROOT OPTION, BE SURE THAT ONLY ONE LINE OF MENU OPTIONS IS DISPLAYED. IF MORE THAN ONE LINE IS DISPLAYED, PRESS **[EXIT]**. NOW, PRESS **[MORE] [F1](MATH)** AND PRESS THE APPROPRIATE F KEY TO SELECT **(ROOT)**. THE CURSOR SHOULD BE PLACED NEAR THE DESIRED ROOT. PRESS **[ENTER]** TO SEE THE DESIRED ROOT WHICH IS THE X-VALUE DISPLAYED AT THE BOTTOM OF THE SCREEN. READ THE TROUBLE SHOOTING NOTE AT THE TOP OF THE NEXT PAGE AND THEN BEGIN THE EXERCISE SET.

To access the ROOT/ZERO option, press **[2nd] <CALC> [2:zero]** (TI-86 users access the root feature as discussed for the TI-85 above and then follow the boxed steps below).

TI-82 TI-82 users are reminded that this option is listed as **[2:root]** and should note that the prompts for lower and upper bounds will appear as "left bound" and "right bound."

STEPS FOR SOLVING AN EQUATION USING THE ROOT/ZERO FEATURE

a. **Set lower bound:** The screen display asks for a lower bound. A lower bound is an X value smaller than the expected root; move the cursor to the left of the left-hand root and press **[ENTER]**. Because the roots are determined on the horizontal axis, a lower bound is always determined by moving the cursor to the <u>left of the root</u> (x-intercept). Notice at the top of the screen a ▸ marker has been placed to designate the location of the lower bound. TI-83/83plus/86 users have the option of not moving the cursor to bound the root, but rather to enter an appropriate value for X at the "lower bound" prompt. Be sure to enter a value for the lower bound that is *smaller* than the expected solution. Press **[ENTER]** after entering the value.

b. **Set upper bound:** Similarly the upper bound is always determined by moving the cursor to the <u>right of the root</u> (x-intercept). At the upper bound prompt, move the cursor to an X value larger than the expected root and press **[ENTER]**. Again, a ◂ marker is at the top of the screen to designate the location of the bound. Again, TI-83/83plus/86 users have the option of not moving the cursor to bound the root, but rather to enter an appropriate value for X at the "upper bound" prompt. Care should be taken that the value entered for an upper bound is *larger* than the expected solution. Press **[ENTER]** after entering the value.

NOTE: If the bound markers do not point toward each other, ▸ ◂, then you will get an "ERROR:bounds" message. If this happens, start the ROOT/ZERO calculation over. Make sure when setting bounds that you bound *only* the specific root for which you are searching.

c. **Locate first root:** Move the cursor to the approximate location of an x-intercept. When you press **[ENTER]** the calculator will search for the root, within the area marked by ▸ and ◂. The root is 0. Make sure you read the "Trouble Shooting Note" at the top of the next page. It addresses round-off error.

d. **Locate subsequent roots:** Repeat the entire process outlined above to determine the right-hand root. This root is 1.

TROUBLE SHOOTING NOTE: There will be times when the calculator will be very close to ZERO but will not display ZERO exactly. For example $X = -7.65 E -15$ is the calculator's version of the scientific notation -7.65×10^{-15} which is equivalent to $X = -.00000000000000765$. For all practical purposes this value is ZERO. To verify that the X value is actually zero, scroll through the TABLE to $X = 0$ (or use EVAL) and note that the Y value is -4, the same as was displayed on the INTERSECT screen. Therefore, when using the graph screen to solve equations/inequalities with the ROOT/ZERO feature, you should be aware that the display coordinate values sometimes approximate the actual mathematical coordinates.

EXERCISE SET

Directions: Solve each of the following quadratics, using the ROOT/ZERO option. So that everyone's graph will look alike, display the graph in the standard viewing window. Sketch your graph display, circle the two roots and record their values in the blanks provided. Beneath each problem, factor the quadratic that you graphed.

1. $X^2 + 8X - 9 = 0$

   The roots are X = _____ and X = _____.

   Factorization: _____

2. $(X-2)^2 + 3X - 10 = 0$

   The roots are X = _____ and X = _____.

   Factorization: _____

3. $X^2 + 6X + 9 = 0$

   The root(s) is/are X = _____ .

   Factorization: _____

   (Note: the point at which the graph touches, but does not cross the X-axis, produces two identical roots - often called a double root.)

✍4.   Compare the factorization of the polynomial to the real roots that were determined in each of the problems above. How do they compare?

5.   If you know that the roots to an equation are 4 and -2, you should be able to write an equation, in factored form that has these roots. Write an equation, in factored form:

   _____

63

6.   The equations that have been solved thus far in this unit have all been second degree equations. Based on the exercises, how many roots should you expect to have with a second degree quadratic equation? _____

✍ 7. **ONE** of the equations does not conform to the pattern. Which one is it and why is the number of roots different from the rest of the problems?

Directions: The next set of equations contain polynomials that are not second degree. However, these polynomials can be factored.
a. First use the ROOT/ZERO method to solve the equation.
b. Copy your screen display.
c. Circle the real roots.
d. Record the value of the roots in fractional form.
e. Factor the polynomial.

8. $X^3 - 7X^2 + 10X = 0$

The roots are X = _____, _____ and _____.

Factorization: _____

9. $X^4 - 5X^2 + 4 = 0$

The roots are X = _____, _____, _____ and _____.

Factorization: _____

10. $3X^4 + 2X^3 - 5X^2 = 0$
Suggestion: Change the WINDOW values (Xmin = -5, Xmax = 5) OR use the **ZBox** option.

The roots are X = _____, _____ and _____.

Note: There is a double root in this problem. Which root is the double root? _____

Factorization: _____

✍11. What conclusions can be drawn about the number of real solutions and the degree of the polynomial equations that have been solved thus far?

NON-FACTORABLE EQUATIONS

Previously, factorable polynomial equations were solved. Several observations were made: 1) for each factor there was a root; 2) the number of factors corresponded to the degree of the polynomial; 3) the number of real roots was equal to or less than the degree of the polynomial. CONCLUSION: A polynomial equation of degree n will have <u>at most n real roots</u>. We will now examine polynomial equations that do not factor over the rational numbers and hence may not have any real roots, or roots that are irrational.

Example 4: Solve the equation $3x^2 - 18x + 25 = 0$.

Solution: This trinomial does not factor, so it would be solved using either the quadratic formula or the method of "completing the square".

a. QUADRATIC FORMULA:	b. COMPLETING THE SQUARE:
$3x^2 - 18x + 25 = 0$	$3x^2 - 18x + 25 = 0$
$a = 3, b = -18, c = 25$	$\frac{1}{3}(3x^2 - 18x + 25) = \frac{1}{3}(0)$
$\dfrac{-b \pm \sqrt{b^2 - 4ac}}{2a} =$	$x^2 - 6x + \dfrac{25}{3} = 0$
	$x^2 - 6x = -\dfrac{25}{3}$
$\dfrac{-(-18) \pm \sqrt{(-18)^2 - 4(3)(25)}}{2(3)} =$	$x^2 - 6x + 9 = -\dfrac{25}{3} + 9$
$\dfrac{18 \pm \sqrt{324 - 300}}{6} =$	$(x - 3)^2 = \dfrac{2}{3}$
	$x - 3 = \pm \sqrt{\dfrac{2}{3}}$
$\dfrac{18 \pm \sqrt{24}}{6} =$	$x = 3 \pm \sqrt{\dfrac{2}{3}}$
$\dfrac{18 \pm 2\sqrt{6}}{6} = \dfrac{9 \pm \sqrt{6}}{3} = 3 \pm \dfrac{\sqrt{6}}{3}$	$x = 3 \pm \dfrac{\sqrt{6}}{3}$

Enter each of the two roots found above into the calculator to determine a decimal approximation. Record the two roots <u>exactly</u> as they are displayed on the screen. DO NOT round.

X = _____ and X = _____

Enter $3X^2 - 18X + 25$ after Y1 = and graph in the standard viewing window. Use the ROOT/ZERO option twice to compute both roots of the equation. The screens displayed indicate both roots. These roots should be comparable to the approximate values that were determined above.

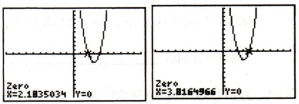

Zero
X=2.1835034 Y=0

Zero
X=3.8164966 Y=0

EXERCISE SET CONT'D

Directions: Use the ROOT/ZERO option to find the **REAL** roots of the following quadratic equations. Sketch the screen display in the indicated viewing window and record the solutions.

12. $-5X^2 + 5X + 8 = 0$
          X = _____ and _____

ZStandard

65

✍13. What happens when you try to convert the roots in #12 to fractions? and why?

14. Solve the quadratic equation in #12 by either completing the square or using the Quadratic Formula. From the home screen, approximate the solutions and compare them to the calculator answers recorded. They should be the same.

✍15. $X^2 + 5X + 8 = 0$

Is it possible to graphically find the roots of this equation using the calculator?_____

Why or why not?

16. Solve #15 analytically using either the Quadratic Formula or by completing the square.

Graphs which intersect the X-axis will have real roots because the X-axis represents the set of real numbers. Roots which are complex numbers will not be represented on the X-axis. Thus an equation with complex roots will not intersect the X-axis. There will, however, be two roots because complex roots always occur in conjugate pairs.

66

17. State the number and type of roots (real or complex) of each of the following equations. DO NOT SOLVE the equations, simply graph the polynomial function in the ZStandard viewing window and check the number of X-intercepts, if any.

a. $6X^2 + 2X - 4 = 0$

Number of roots:_____

Type of roots:_____

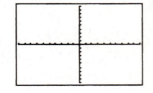

b. $2X^3 - 5X + 5 = 0$

Number of roots:_____

Type of roots:_____

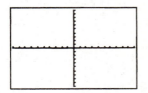

c. $X^4 - 0.5X^3 - 5X^2 + 10 = 0$

Number of roots:_____

Type of roots:_____

18. $0 = X^6 + 2X^3 - 1$
(Zdecimal is a preset window under your Zoom menu)

X = _____ and _____

Describe the nature of the other four roots:

ZDecimal

19. $0 = X^4 - 2X^3 + X - 2$

X = _____ and _____

The expression $X^4 - 2X^3 + X - 2$ in completely factored form is $(X - 2)(X + 1)(X^2 - X + 1)$. Use the appropriate algebraic technique to determine the complete solution set.

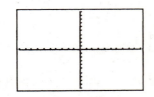

ZStandard

✍20. IN YOUR OWN WORDS:
a. Explain how to use the calculator to find real roots of quadratic equations (you should address both the INTERSECT and ROOT/ZERO feature).
b. The relationship between the graph of the quadratic and the number and type of roots.

Solutions: 1. $(X+9)(X-1)=0$, $X=-9$ and $X=1$ **2.** $(X-3)(X+2)=0$, $X=-2$ and $X=3$,

3. $(X+3)(X+3)=0$, $X=-3$ **4.** The factor is always the "variable minus the root".

5. $(X-4)(X+2)=0$ **6.** two **7.** #3. We did not see two distinct real roots, but rather two identical roots (often called a double root).

8. $X(X-5)(X-2)=0$, $X=0,2$ and 5   **9.** $(X-2)(X+2)(X-1)(X+1)=0$, $X=2,-2,1$ and -1

10. $X^2(3X+5)(X-1)=0$, $X=0,-5/3$ and 1, Zero is the double root. **11.** The number of

solutions is the same as the degree of the equation. However, not all the solutions are

distinct (different). Some appear as multiple roots. **12.** $X=-.860147$ and $X=1.8601471$

13. These answers will not convert to fractions because they are the decimal approximations of the irrational numbers $\frac{1}{2} \pm \frac{\sqrt{185}}{10}$. **15.** No, It does not intersect the X-axis and therefore has no real roots.

16. $X = \frac{-5 \pm i\sqrt{7}}{2}$ **17. a.** 2, real **b.** 3, 1 real, 2 complex **c.** 0, 4 complex

18. $X = -1.341504$, $X = .74543212$; the remaining four roots consist of 2 pairs of complex

conjugates **19.** $X = -2$, $X = 1$, $\left\{2, -1, \frac{1 \pm i\sqrt{3}}{2}\right\}$

UNIT 10
APPLICATIONS OF QUADRATIC EQUATIONS

*Unit 9 is a prerequisite for this unit. Answers appear at the end of the unit.

This unit explores the various features of the calculator that can be used to investigate application problems.

1.   A traveling circus has a "human cannonball" act as its grand finale. The equation $Y = -.01X^2 + .64X + 9.76$ models the flight path of the human cannonball. The variable Y represents the height in feet and the variable X represents the horizontal distance traveled in feet. Display a graphical representation of this equation and use the TRACE feature to answer the questions.

a. TRACE along the graph in the standard viewing window and examine the numbers (x and y-coordinates) displayed at the bottom of the screen.

What do the X values represent and what do the Y values represent - within the context of the problem?

b. Are these decimal representations appropriate within the context of the problem? Why or why not?

In order to have "friendly" values for X and Y, change the viewing WINDOW to ZInteger.

TI-85/86	TI-85/86 USERS PRESS **[GRAPH] [F3](ZOOM) [MORE] [MORE]** AND THE APPROPRIATE F KEY FOR **(ZINT)**.

Be sure your cursor is at X = 0 and Y = 0 (use the arrow keys to move the cursor to the point where the axes intersect to display X = 0 and Y = 0). Press **[ZOOM]**, cursor down to highlight 8 for ZInteger and press **[ENTER]** to choose this option. Press **[ENTER]** again to set the viewing WINDOW to ZInteger. The blinking cursor was positioned at X = 0 and Y = 0 to keep the axes centered on the screen when changing to ZInteger. The cursor may be placed at any position on the screen and the axes will intersect at that point. Now TRACE along the curve and observe the "friendly" values represented at the bottom of the screen.

NOTE: ZInteger yields integer values for X when tracing on the graph. This is particularly valuable when X only has meaning as an integer value.

c. This curve represents the flight path of the human cannonball. **TRACE** along the path to the right. How far, <u>approximately</u>, has the human cannonball traveled horizontally when he hits the ground?_____

d. The human cannonball is traveling at speeds up to 65 mph. To land on the ground would mean certain death. If he uses a net for his landing, how far will he have traveled horizontally if the net is 11 feet above the ground?_____

e. What is the maximum height reached during the course of his flight?_____

f. How far has he traveled horizontally when he reaches this maximum height?_____

NOTE: The actual distance he has traveled is a length of arc along the curve of the parabola. To calculate this length requires the use of calculus.

g. The human cannonball is shot out of the cannon head first, so all of the distances are measured from his head. TRACE along the curve to X = 0 and Y = 9.76. Explain the meaning of these two values.

2. Blaire is a pitcher for the Girls Slowpitch softball team at her middle school. The height of the softball X seconds after she releases a pitch is given by the formula $h = -16X^2 + 18X + 3$.

a. Find the length of time it will take the ball to hit the ground if the batter swings and misses.

Solution: When the ball hits the ground the height will be h = _____. Thus the equation we are trying to solve for X is $0 = -16X^2 + 18X + 3$. We want to solve this equation using the root/zero feature, so enter $-16X^2 + 18X + 3$ after Y1 = and set the standard viewing window. You will need to adjust the viewing rectangle!

Press [**TRACE**] to determine the points it crosses the X-axis (it appears to cross at about - 0.14 and 1.26) and to determine the approximate height (the highest y-value is about 8). Based on this information, press [**WINDOW**] and change the Xmin to - 0.2, Xmax to 1.5, Ymin to - 5 and Ymax to 10. Press [**TRACE**] to simultaneously view the graph and to activate the

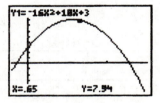

TRACE feature. TRACE around the curve to try to determine the exact time the ball will be 0 feet from the ground. TRACE will not yield the exact answer. Now use the ROOT/ZERO option to determine the two roots/zeroes of the equation.

X = _____   or X = _____

One of these roots is not valid. Which one is it and why?

If you are having difficulty answering "why" then ask yourself this question: Can X equal a negative number? Remember, X represents time.

After using the ROOT/ZERO option on the valid root, you should have determined X to be approximately 1.2723635. This means that it will take the ball approximately 1.27 seconds to hit the ground if the batter misses.

b. What is the highest point that the ball reaches during the pitch?

Solution: Again, TRACE can be used to find the highest point, but letting the calculator determine the MAXIMUM point will be more accurate. Press [**2nd**]

<CALC> [4:maximum]. The calculator is now ready to determine the maximum point on the curve. Set upper and lower bounds as in the ROOT/ZERO option.

Remember, the X value displayed represents seconds elapsed since the pitch was thrown and the Y value indicates the height of the ball at that point in time. Y = ___ indicates that the ball reached a maximum height of approximately _____ feet.

c. How **long** does it take the ball to reach its maximum height? _____

d. What is the height of the ball 1.4 seconds after the pitcher releases it? **(TRACE** along the path of the curve to X = 1.4 seconds.) Carefully explain your answer.

e. Use the TRACE cursor to TRACE along the path of the curve from left to right. Explain, in your own words, what information the X and Y values at the bottom of the screen are giving you.

f. Does the curve represent the path of the ball in flight? If you answer yes, then explain which part of the graph display represents the distance the ball travels.

If you answer no, explain why not.

3.   Bridges are often supported by arches in the shape of a parabola. The equation

$$Y = \frac{10}{7}X - \frac{2}{49}X^2$$, where Y = height and X = distance from the base of the arch,

provides a model for a specific parabolic arch that supports a bridge. Will this arch be tall enough for a road crew to build a county road under?

Solution:
Setting the viewing WINDOW
a. Begin by entering the polynomial at the Y1 = prompt and graph in the standard viewing rectangle.

b. This curve represents the support to a bridge. The entire curve should be visible. TRACE along the curve, recording the following (to the nearest integer): left most X-intercept, maximum Y value, and right most X-intercept.

left most X-intercept: _____   maximum Y value: _____

right most X-intercept: _____

c.	Use the information from part b to set the WINDOW values so that the entire graph is displayed. Press **[WINDOW]** and set the WINDOW values as follows: Xmin = -1, Xmax = 37, Xscl = 1, Ymin = -1, Ymax = 13, Yscl = 1.These values were selected to ensure that the area slightly beyond the perimeter of the graph is displayed.

NOTE: For more information on setting viewing WINDOWS, refer to the unit entitled: "Where Did the Graph Go?".

Solving the problem:

d.	Begin by determining the height (to the nearest tenth) of the highest point under the arch.

	Maximum height = _____

✍e.	If the average vehicle is no more than 6 feet high, can the vehicle drive under the arch?_____ Explain how you determined your answer.

✍f.	If a two lane road is 20 feet wide, will it fit between the bases of the arch?_____ Explain how you determined your answer.

✍g.	Can the average vehicle drive in either lane under the arch and not scrape the paint off the roof? (or scrape the roof off the car??) That is to say, if this 20 foot wide road is centered under the arch, is the arch at least 6 feet above the road at all points in its width?

4.	The local community theater is considering increasing the price of its tickets to cover increases in costuming and stage effects. They must be careful because a ticket price that is too low will mean that expenses are not covered and yet a ticket price that is too high will discourage people from attending. They estimate the total profit, Y, by the formula $Y = -X^2 + 35X - 150$, where X is the cost of the ticket.

Before attempting to graphically solve the problem, set the viewing WINDOW by following steps a - c in #3. **MAKE SURE THAT THE ENTIRE CURVE IS DISPLAYED.**

a. What is the maximum amount that can be charged for a ticket to maximize the profit?_____

b. What is the maximum profit?_____

c. If $13 dollars is charged for each ticket, what will the profit be?
What are your solution options here? We could return to the home screen and evaluate the polynomial for X = 13 by using the **STO**re feature or we could scroll through the TABLE in search of X = 13. However, since we have been using the CALC menu to investigate the graph, we will look at value (EVAL X) which is the first entry option under the CALC menu. Access the CALC menu and press **[1:value]** to select value and display the **X** prompt (displayed as EVAL X on the TI-82/85/86). At the prompt, enter 13 and press **[ENTER]**. What will the profit be when $13 is charged for each ticket? _____

TI-85/86	RECALL THAT **EVAL X** IS ACCESSED BY PRESSING **[MORE]** TWICE (THE GRAPH MENU MUST BE DISPLAYED) AND SELECTING **"EVAL."**

BEWARE: The value (EVAL) feature only works when the value selected for X is between the Xmax and Xmin values on the WINDOW screen.

d. If they predict a profit of $150.00 on a play, how much was charged per ticket? (Scroll through the Y values in the TABLE to answer this question.)_____

e. Use the TABLE display to determine at what point the theater "breaks even", i.e. How much must each ticket cost for there to be no money lost and yet no profit made?

✍5. IN YOUR OWN WORDS: Write a summary of what you learned in this unit. You should address the following:
a. how and when to adjust WINDOW values,
b. use of the MAXIMUM option, and
c. use of the value (EVAL X) option, including its restrictions.

Solutions: **1c.** approx. 76.5 ft. **1d.** 62 ft. **1e.** 20 ft. **1f.** 32 ft. **1g.** It means that his head is 9.76 feet above the ground before he is shot from the cannon. **2a.** h = 0, x = -.1473635, x = 1.2723635 **2b.** Y = 8.0625, 8 feet **2c.** approx. six tenths of a second **2d.** Y = -3.42, the ball is 3.41 feet into the ground. **2e.** The X values represent the time (in seconds) that the ball is in the air; the Y values represent its height. **2f.** Yes: This is obviously an incorrect answer, because nothing represents distance. No: This is the correct response, because the graph is relating the time (X) to the height of the ball (Y). **3d.** 12.5 **3e.** Yes **3f.** Yes, because the supports are 35 feet apart. **3g.** Yes **4a.** $17.50 **4b.** $156.25 **4c.** $136 **4d.** 15, 20 **4e.** The theater breaks even when tickets are priced at $5 each or $30 each.

UNIT 11
GRAPHICAL SOLUTIONS: RADICAL EQUATIONS

*Unit 9 is a prerequisite for this unit. Answers appear at the end of the unit.

This unit will investigate using the use of the INTERSECT and ROOT/ZERO options to solve equations that contain radicals. Recall, the solution to an equation is the value(s) for the variable that produce a true arithmetic statement.

Solve $\sqrt{3X + 7} + 2 = 7$ algebraically   Check your solution(s) by substitution:

To graphically solve this equation we will first look at the graphical representation of each side of the equation. Enter $\sqrt{3X + 7} + 2$ at Y1 = and the constant 7 at Y2 =. Graph in the standard viewing window and compare your graph to the screen pictured at the right. Recall, we want to determine graphically where Y1 = Y2. Circle the point of intersection.

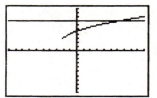

Use the INTERSECT option to graphically find the X-coordinate of the intersection of the two graphs. The X-coordinate of the intersection should be the same as the value found algebraically above: 6. If it is not, recheck both the algebraic solution and the calculator solution.

Now use the ROOT/ZERO option to graphically solve the same radical equation. Remember, you must first rewrite the equation with all terms on one side of the equal sign and the other side equal to 0. Do this in the space below.

If the instructor does not require you to show this algebraic computation, then turn "off" the graphs of Y1 and Y2 and merely enter Y1-Y2 at the Y3 = prompt. (See OPTIONAL NOTE in the middle of page 61.)

Press **[GRAPH]** and compare your screen to the one pictured at the right. Circle the root/zero, i.e. the X-intercept.

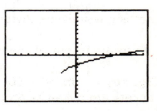

Use the ROOT/ZERO option to find the root of the equation. Your root should, of course, be 6.

Directions: Use either the ROOT/ZERO or the INTERSECT options on the calculator to solve each radical equation below. Sketch the screen display and use the ▸Frac option (under the MATH menu) to convert all decimal results to fractions.

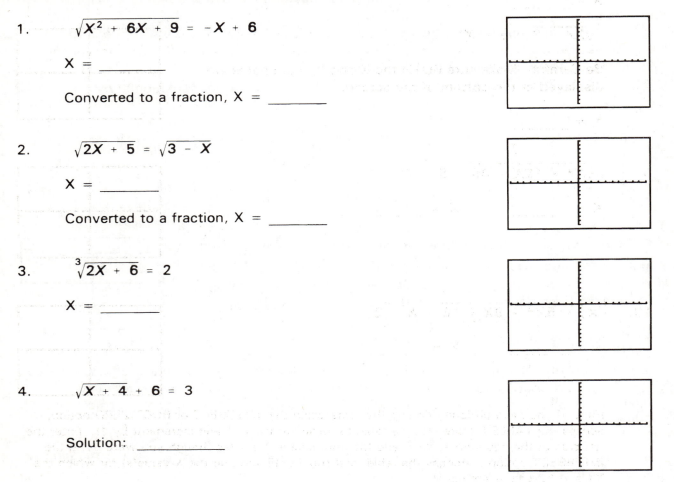

1. $\sqrt{X^2 + 6X + 9} = -X + 6$

    X = _____

    Converted to a fraction, X = _____

2. $\sqrt{2X + 5} = \sqrt{3 - X}$

    X = _____

    Converted to a fraction, X = _____

3. $\sqrt[3]{2X + 6} = 2$

    X = _____

4. $\sqrt{X + 4} + 6 = 3$

    Solution: _____

✍ 5. If you solved the equation in #4 algebraically, the first step would be to isolate the radical. Once the radical is isolated, you should realize there are no solutions. Why?

6. Solve the equation $\sqrt{X} + 2 = \sqrt{5 - X} + 3$ algebraically in the space below. Check your solution(s) by substitution.

    Algebraic Solution   Check by substitution

7.   Graph the equation in #6 by entering the left side at Y1 = and the right side at Y2 = . Copy the display screen. How many points of intersection do you see? Use the calculator to find the solution.

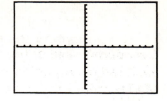

X = _____

8. $\sqrt{2X - 3} = 3 - X$

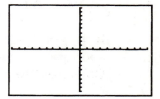

Be Careful! Make sure BOTH the X and Y coordinates are displayed at the bottom of the screen.

X = _____

9. $\sqrt{X^2 - 12X + 36} + 5 = 7$

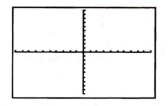

X = _____   X = _____

10. $\sqrt[3]{X^3 + 5X^2 + 9X + 18} = X + 2$

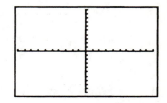

X = _____   X = _____

Hint: If you have difficulty finding the roots using the INTERSECT or ROOT/ZERO option, access the TABLE feature (set the table to begin with X = 1 and increment by 1). Enter the left side of the equation at Y1 = and the right side at Y2 = (as though you were using the INTERSECT option). Access the table, and scroll until you find the X-value(s) for which the Y1 and Y2 values are equal.

11. In the space below, solve $\sqrt{-2X + 6} = 3 - X$ algebraically. Check solutions by substitution.

Algebraic solution: Check by substitution:

12. a. Use the INTERSECT option and have the calculator find the
     solutions to the equation in #11. Copy your screen display.

     X = _____

     b. Use the ROOT/ZERO option and have the calculator find the
     solutions. Copy your screen display.

     X = _____

     ✍ c. You found two valid roots algebraically, two roots were displayed graphically,
     and yet only one root could be computed with the calculator. Which is the correct
     solution - the algebraic solution in #11 or the calculator solutions above?

     d. Set your table to a start value of 1 and increment by 1. Make sure you have the
     left side of the equation entered at Y1 = and the right side at Y2 =. Access the
     table. For what values of X are the Y1 and Y2 values equal?

     X = _____   X = _____ This confirms your algebraic solution.

     ✍ e. Explain **why** the calculator was unable to compute both roots in part a.

13. **Application:** The period of a pendulum on a clock is the time
     required for the pendulum to complete one cycle (one "swing"
     from a given position back to this initial point). The formula for
     finding the period of a pendulum is $T = 2\pi\sqrt{\dfrac{X}{32}}$, where T is the

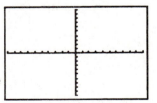

     time required in seconds and X is the length of the pendulum.
     A clock company is constructing a clock for a window display. If it takes the
     pendulum two seconds to complete 1 period, what is the length of the pendulum (to
     the nearest hundredth of a foot)?

NOTE: Work problems from your text using what you have learned in this unit. Decide what works best for <u>you</u> - algebraic solutions? the ROOT/ZERO option? the INTERSECT option? Then PRACTICE.

✍14. **IN YOUR OWN WORDS:** Summarize what you learned in this unit. Your summary should address
  a. the use of the INTERSECT option to solve radical equations,
  b. the use of the ROOT/ZERO option to solve radical equations,
  c. the use of the TABLE feature to find roots.

Solutions: **1.**. X = 1.5, 3/2 **2.** X = -.6666667, -2/3 **3.** 1 **4.** Null Set

5. Because the right side would be equal to -3 and $\sqrt{}$ is defined only for positive roots.

6. X = 4, One is an extraneous root. **7.** X = 4 **8.** X = 2 **9.** X = 4 or X = 8

10. X = -5 or X = 2 **11.** X = 1 or X = 3

12a. X = 1 **12b.** X = 1 **12c.** {1, 3} - Both of these solutions check algebraically.

12d. X = 1 or X = 3

12e. When using the INTERSECT feature, the calculator establishes upper and lower bounds using the domain of the graphed functions. It then searches between these bounds for the point of intersection, **excluding the bounds in its search**.

13. 3.24 feet

UNIT 12
GRAPHICAL SOLUTIONS: LINEAR INEQUALITIES

*Unit 7 is a prerequisite for this unit. Answers appear at the end of the unit.

To solve linear inequalities such as 5X - 1 $\geq$ -3, we want to find replacement values for X that will produce a true arithmetic sentence. The solution to a first degree inequality in one variable is typically an infinite set of numbers rather than a single number.
  Solve 5X - 1 $\geq$ -3, algebraically:

The -2/5 you got in your solution is a "critical point." It is so named because it divides the number line into three distinct subsets of numbers - those larger than the number, those smaller than the number, and the number itself.

  In the space below, test the critical point -2/5 in the original inequality.
  (i.e. When X is replaced by -2/5, is the resulting inequality a true statement?)

  In the space below, test a number whose value is larger than that of the critical point.

Because the test number tested true, the implication is that all values to the right of the critical point will also test true. This is also indicated by the mathematical statement X $\geq$ -2/5, the algebraic solution.

What *should* happen when a number is tested whose value is smaller than that of the critical point? If you are not sure, choose a value and test it!

The graphing calculator can be used to quickly confirm the solution of X $\geq$ -0.4.
Press [Y =] and enter 5X - 1 after Y1 = and -3 after Y2 =. Examine the TABLE values where X = -0.4, X > -0.4 and X < -0.4. To do this, press [2nd] <TBLSET> and set the table to start at -.4 and increment by 1. (TI-86 users press [TABLE] and then [F1](TBLST).) Press [2nd] <TABLE> (TI-86 users press F2) to view the table of values. Compare the Y1 and Y2 columns for X = -0.4, X > -0.4 and X < -0.4. (Recall, Y1 = 5x - 1 and Y2 = -3.) Your table should correspond to the one at the right. When X = -0.4, the expression 5X -1 is equal to -3. When X > -0.4, the expression 5X - 1 has values greater than -3 and when X < -0.4, the expression 5X - 1 has values less than -3.

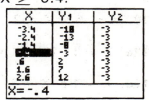

Now consider the graphical solution. Since $Y1 = 5X - 1$ and $Y2 = -3$, we want to locate graphically <u>where</u> $Y1 \geq Y2$.

Press [ZOOM] [6:ZStandard] to set the standard viewing WINDOW. The WINDOW values at the right are automatically entered and the graph screen will be displayed. Pressing [WINDOW] displays these values.

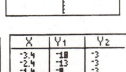

```
WINDOW
 Xmin=-10
 Xmax=10
 Xscl=1
 Ymin=-10
 Ymax=10
 Yscl=1
 Xres=1
```

TI-85/86 TI-85/86 USERS PRESS **[GRAPH] [F3](ZOOM)** AND **(ZSTD)** TO SET THE STANDARD VIEWING WINDOW.

Press [GRAPH] and sketch the graphical display of the inequality, being careful to label "Y1" and "Y2" as their graphs are displayed. Circle the point where the two graphs are equal, the point of intersection. Use the INTERSECT option to determine the solution to the equation $5X - 1 = -3$. You should get $X = -0.4$.

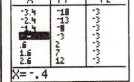

Graphs should be read from left to right. The graph is a visual representation of numerical information that was previously displayed in the TABLE. For what X values is Y1 greater than Y2? It should be the section that is highlighted on the graph displayed. In general, $Y1 > Y2$ where the graph of Y1 is <u>above</u> the graph of Y2.

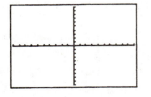

The solution to the inequality is $X \geq -0.4$.

In set notation this would be written: $\{X \mid X \geq -0.4\}$

As a number line graph this would be:

In interval notation this would be written $[-0.4, \infty)$.

EXERCISE SET

1. **Solve** $10 - 3X < 2X + 5$ graphically.

   Solution Steps:
   a. Press [Y =] and enter $10 - 3X$ after $Y1 =$ and $2X + 5$ after $Y2 =$. Sketch the graph displayed. Be observant the first time the graphs are displayed. It will be helpful to label Y1 and Y2. The graph of Y1 will appear first.

   b. Use the INTERSECT feature to find the point of intersection. The intersection is at $X = \underline{\qquad}$. The expressions entered at Y1 and Y2 are equivalent when $X = 1$. Therefore, $X = 1$ would be the solution to the *equation* $10 - 3X = 2X + 5$.

   c. Use a highlighter pen to highlight the section of Y1 that is **less than** Y2. Recall, $Y1 < Y2$ when the graph of Y1 is <u>below</u> the graph of Y2.

d. Determine your solution by setting **TblStart** = *critical point*. Examine the relationship between the values of Y1 and Y2 for X values that are both greater than, equal to, and less than the critical point.

e. Conclusion: Y1 < Y2 when X > 1.
The solution set is { X |X > 1}.
Translate this solution to a number line graph:

2. **Solve** 6 - 5X $\leq$ -1 graphically, following the steps outlined in 1.

   Solution Set:_____

   Translate this solution to a number line graph:

3. **Solve** -3 $\geq$ 7 - 2X graphically, following the steps outlined in 1.

   Solution Set:_____

   Translate this solution to a number line graph:

4. **Solve** $\dfrac{4X - 2}{6} < \dfrac{2(4 - X)}{3}$ graphically, following the steps outlined in 1.

   Solution Set:_____

   Translate this solution to a number line graph:

5. A linear inequality has been entered on the calculator so that the left side is expressed as Y1 and the right side as Y2.
   a. What is the critical point as displayed in the TABLE of values?____

X	Y1	Y2
-2	0	12
-1	2	11
0	4	10
1	6	9
2	8	8
3	10	7
4	12	6

   X= -2

   b. What is the solution (expressed as a number line graph) to the inequality Y1 < Y2?

Directions: **Solve** each inequality graphically using the INTERSECT feature. Begin in a ZStandard WINDOW and adjust the viewing window accordingly. Critical points should be expressed in fraction form.

6. $3X - 8 < \dfrac{4}{5}(3X - 2)$

Xmin = _____   Xmax = _____

Ymin = _____   Ymax = _____

Solution Set: _____   Interval Notation: _____

Translate this solution to a number line graph:

7. $8 + \dfrac{8}{5}X \geq \dfrac{1 + 9X}{8}$

Xmin = _____   Xmax = _____

Ymin = _____   Ymax = _____

Solution Set: _____   Interval Notation: _____

Translate this solution to a number line graph:

8. $-\sqrt{288} + X > 2(2X - 6\sqrt{2})$

Xmin = _____   Xmax = _____

Ymin = _____   Ymax = _____

Solution Set: _____   Interval Notation: _____

Translate this solution to a number line graph:

✍9. Solve the combined inequality $-4 < 4 + 2x < 8$ graphically. Use the steps outlined in Exercise 1 as a model. Record any additional steps required to obtain a valid solution. Display your graph at the right.

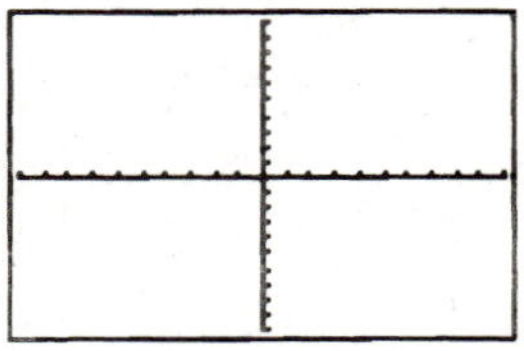

Solution Set: _____

Translate this solution to a number line graph:

10. The graph displayed at the right is the graph of the linear equation Y = P where P is a linear polynomial expression. The coordinates displayed represent the point where the line intersects the X-axis (X-intercept). Solve the indicated inequalities using the displayed graph and record your solution as a solution set and as a number line graph on the number line provided.

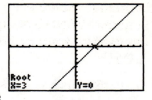

a. Y = 0   Solution set: _____

b. Y > 0   Solution set: _____

c. Y < 0   Solution set: _____

d. Y ≤ 0   Solution set: _____

e. Y ≥ 0   Solution set: _____

✍11. **IN YOUR OWN WORDS:** Summarize what you have learned in this unit about graphically solving linear inequalities. Be sure to include:
a. a comparison of solving an <u>equation</u> versus solving an <u>inequality</u>,
b. how to determine the critical point and whether or not it is included in the solution, and
c. how to decide whether to include the numbers greater than the critical point in the solution set or those numbers that are less than the critical point.

THE FOLLOWING PROVIDES INFORMATION ABOUT THE GRAPH STYLE ICON (APPLICABLE ONLY TO THE TI-83/83PLUS AND TI-86) WHICH YOU MAY WANT TO EXPERIMENT WITH WHEN GRAPHING MORE THAN ONE POLYNOMIAL AT A TIME. IT ALLOWS YOU TO DISTIINQUISH BETWEEN THE GRAPHS OF EQUATIONS.

TI-83/83plus

The "graph style icon" is a feature that allows you to distinguish between the graphs of equations. Press **[Y=]** and observe the "\" in front of each Y. Use the left arrow to cursor over to this "\". The diagonal should now be moving up and down. Pressing **[ENTER]** once changes the diagonal from thin to thick. The graph of Y1 will now be displayed as a thick line. Repeatedly pressing **[ENTER]** displays the following:

- ▀: shades above Y1
- ▄: shades below Y1
- -o: traces the leading edge of the graph followed by the graph
- o: traces the path but does not plot
- ·.: displays graph in dot, not connected MODE

The graphing icon takes precedence over the **MODE** screen. If the icon is set for a solid line and the **MODE** screen is set for DOT and not solid, the graphing icon will determine how the graph is displayed. Pressing **[CLEAR]** to delete an entry at the Y= prompt will automatically reset the graphing icon to default, a solid line.

TI-86

THE "GRAPH STYLE ICON" ALLOWS YOU TO DISTINGUISH BETWEEN THE GRAPHS OF EQUATIONS. PRESS **[GRAPH] [F1](Y(X) =)** AND OBSERVE THE "\" IN FRONT OF EACH Y. PRESS **[MORE]** FOLLOWED BY **[F3](STYLE)**. THE DIAGONAL SHOULD NOW HAVE CHANGED FROM THIN TO THICK. THE GRAPH OF Y1 WILL BE DISPLAYED AS A THICK LINE. REPEATEDLY PRESS **[STYLE]** TO DISPLAY THE FOLLOWING:

- ▀: SHADES ABOVE Y1
- ▄: SHADES BELOW Y1
- -O: TRACES THE LEADING EDGE OF THE GRAPH FOLLOWED BY THE GRAPH
- O: TRACES THE PATH BUT DOES NOT PLOT
- ·.: DISPLAYS GRAPH IN DOT, NOT CONNECTED MODE

IT IS IMPORTANT TO REMEMBER THAT THE GRAPHING ICON TAKES PRECEDENCE OVER THE **MODE** SCREEN. IF THE ICON IS SET FOR A SOLID LINE AND THE **MODE** SCREEN IS SET FOR DOT AND NOT SOLID, THE GRAPHING ICON WILL DETERMINE HOW THE GRAPH IS DISPLAYED. PRESSING **[CLEAR]** TO DELETE AN ENTRY AT THE Y = PROMPT WILL AUTOMATICALLY RESET THE GRAPHING ICON TO DEFAULT, A SOLID LINE.

Solutions to Exercise Sets: 1. $\{x \mid x > 1\}$ 2. $\{X \mid X \geq 1.4\}$

3. $\{X \mid X \geq 5\}$ 3. $\{X \mid X < 2.25\}$

5. The critical point is 2 and the number line graph of the solution is

6. $\{X \mid X < 32/3\}$, $(-\infty, 32/3)$

7. $\{X \mid X \geq -315/19\}$, $[-315/19, \infty)$

8. $\{X \mid X < 0\}$, $(-\infty, 0)$

9. $\{X \mid -4 < X < 2\}$ or (in interval notation) $(-4, 2)$

10. **a.** $\{3\}$ **b.** $\{X \mid X > 3\}$ **c.** $\{X \mid X < 3\}$ **d.** $\{X \mid X \leq 3\}$ **e.** $\{X \mid X \geq 3\}$

UNIT 13
GRAPHICAL SOLUTIONS: ABSOLUTE VALUE INEQUALITIES

*Unit 12 is a prerequisite for this unit. Answers appear at the end of the unit.

In the prerequisite unit, linear inequalities were solved graphically using the INTERSECT feature aided by the TABLE. The same approach will be used to solve inequalities containing absolute value.

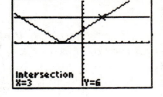

Consider $|X + 3| \le 6$. The inequality translates to two statements: find the absolute value of the quantity X plus three less than 6 ($|X + 3| < 6$) or find the absolute value of the quantity X plus three equal to 6 ($|X + 3| = 6$). The INTERSECT feature can be used to quickly determine that the solutions to the equation are 3 and - 9. Label Y1 and Y2 on the graph at the right (83/83plus/86 users may want to try the graph style icon and graph Y2 as a thick line instead of using a label).

Look at the displayed graph. We know that Y1 < Y2 when the graph of Y1 is **below** the graph of Y2. To answer the second part of the question, we must determine all X values for which this is true. We must find the values of X that produce the highlighted portion of the pictured graphical display.

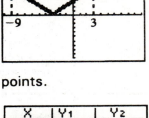

On the graph, we drew a dotted vertical line from each point of intersection perpendicular to the horizontal axis. We counted the tic marks on the horizontal axis and labeled the points where the perpendicular lines touched the x-axis.

The highlighted portion of the graph of Y1 is between the two labeled points.

Verify this with your TABLE. Press [2ⁿᵈ] <TBLSET> (TI-86 users press [**TABLE**] [**F2**] (**TBLST**)) and start the table at -12 with an increment of 3. Compare the Y1 and Y2 columns for X < -9, -9 < X < 3, and X > 3. Recall when Y1 = 6 that X = -9 or X = 3. The TABLE verifies that Y! ≥ Y2 when -9 ≤ X ≤ 3.

X	Y1	Y2
-12	9	6
-9	6	6
-6	3	6
-3	0	6
0	3	6
3	6	6
6	9	6

X= -12

The solutions to the **inequality** $|X + 3| < 6$ are all the X values between -9 and 3 (the critical points).

The solutions to the **equation** $|X + 3)| = 6$ are the two critical points -9 and 3.

Combining this information, we get the solution to $|X - (-3)| \le 6$ to be $-9 \le X \le 3$.

Solution Set: $\{X | -9 \le X \le 3\}$, Number line graph:

Interval Notation: [-9,3]

87

Graphically solve $|X + 3| > 6$ by following the indicated steps.

a. Press [**Y =**] and enter abs(X + 3) after Y1 = and 6 after Y2 = .

b. Press [**GRAPH**]. Sketch the display screen and draw dotted lines from the intersection points perpendicularly to the horizontal axis.

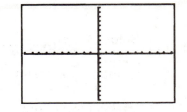

c. Use the calculator's INTERSECT option to find the points of intersection.
X = _____ and X = _____

d. Count the tic marks and label the points on the horizontal axis where your perpendicular lines touch the axis.

NOTE: At this point, the steps listed above are the same steps you used to solve $|X + 3| \leq 6$ in the first part of this unit and to solve $|X + 3| = 6$ in a previous unit.

e. The solution is found to be the X values where Y1 > Y2. When the graph of Y1 is above the graph of Y2, as indicated by the highlighted portions, then Y1 > Y2. The TABLE display supports the graphical interpretation.

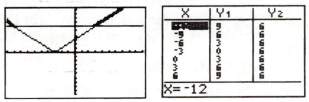

f. The solutions to the inequality $|X + 3| > 6$ are all X values greater than 3 or less than -9: X > 3 or X < -9.

Solution Set: {X | X < -9 or X > 3}

Number line graph:

Interval Notation: (-∞,-9) ∪ (3,∞)

NOTE: *-9 and 3 were not included because the inequality states that $|X + 3|$ is* <u>*strictly*</u> *greater than 6.*

EXERCISE SET

Directions: Graphically solve each of the following inequalities following the steps that were outlined. Record the solution in set notation, as a number line graph and in interval notation.

1. $|2X - 1| \geq 5$

Solution Set:_____

Number line graph:

Interval Notation:_____

2. $\left|\dfrac{1}{2}X - 1\right| < 4$

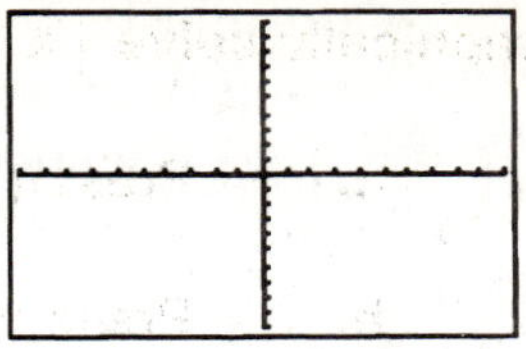

(Be Careful! Enclose the $\dfrac{1}{2}$ in parentheses.)

Solution Set:_____

Number line graph:

Interval Notation:_____

3. $\left|\dfrac{2X + 5}{3}\right| < 4$

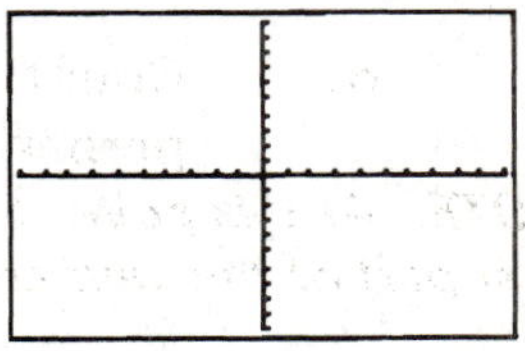

Solution Set:_____

Number line graph:

Interval Notation:_____

✍4. In your own words, explain what type of error could easily be made when graphing the expression $\left|\dfrac{2X + 5}{3}\right|$ or $\left|\dfrac{1}{2}X - 1\right|$.

5. $\left|4X + 2\right| > -3$
This particular inequality represents a "special case." Carefully re-TRACE your highlighted portion of the graph before deciding on the solution.

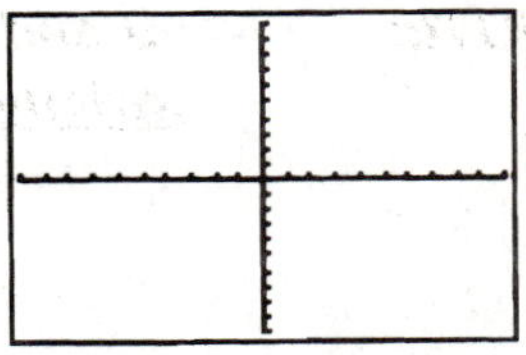

Solution Set:_____

Number line graph:

Interval Notation:_____

6. $\left|4X + 2\right| < -3$ (another "special case")

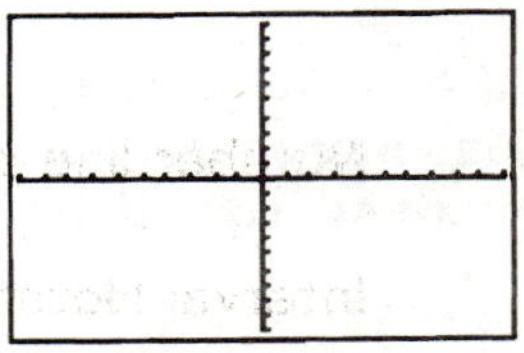

Solution Set:_____

✍7.	Consider why #5 and #6 are labeled as "special cases." Could #5 and #6 have been solved by merely "looking" at the inequality? Think carefully about the definition of absolute value before formulating your response.

✍8.	IN YOUR WORDS: Because the left and the right sides of equations and inequalities are graphed as separate expressions, the graphical representations of the solutions to each of the following problems all look alike.

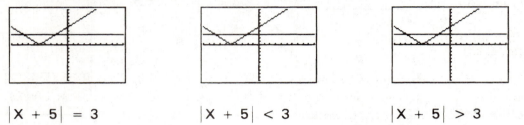

$|X + 5| = 3$ $|X + 5| < 3$ $|X + 5| > 3$

To summarize your results from this unit, explain how to interpret the solution represented by each graphical display.

a. Interpretation of $|X + 5| = 3$:

b. Interpretation of $|X + 5| < 3$:

c. Interpretation of $|X + 5| > 3$:

Solutions: 1. $\{X \mid X \leq -2 \text{ or } X \geq 3\}$,   , $(-\infty, -2] \cup [3, \infty)$
$$\overset{\bullet}{\underset{-2}{\qquad}}\qquad\overset{\bullet}{\underset{3}{\qquad}}$$

2. $\{X \mid -6 < X < 10\}$, $(-6, 10)$,
$$\overset{\circ}{\underset{-6}{\qquad}}\qquad\overset{\circ}{\underset{10}{\qquad}}$$

3. $\{X \mid -8.5 < X < 3.5\}$,   , $(-8.5, 3.5)$
$$\overset{\circ}{\underset{-8.5}{\qquad}}\qquad\overset{\circ}{\underset{3.5}{\qquad}}$$

4. You might not put parentheses around the numerator of the fractions or fail to enclose the entire fraction in parentheses when using the absolute value command.

5. $\mathbb{R}$ or $\{X \mid X \text{ is a real number}\}$, $(-\infty, \infty)$ 6. null set

7. Remember that an absolute value is at least 0 or larger. Thus, an absolute value is always greater than a negative number for any value of the variable and an absolute is never less than any negative number regardless of the value of the variable.

8. The graph displays all look the same because we are always entering the left side of the equation/inequality at Y1 and the right side at Y2. It is the interpretation of these graphs that yields the correct solution.

UNIT 14
GRAPHICAL SOLUTIONS: QUADRATIC INEQUALITIES

*Unit 9 is a prerequisite for this unit. Answers appear at the end of the unit.

This unit will graphically examine quadratic inequalities by using the ROOT/ZERO option of the calculator and by interpreting the relationship between the graphical displays of each side of the inequality. REMEMBER: To use the ROOT/ZERO there must be a zero on one side of the inequality.

SPECIAL CASES

The first four examples represent "special cases." They will be the quickest to solve of all the inequalities. All graphs will be displayed in the standard viewing window unless otherwise noted. Set this viewing window now.

Example 1: Graphically solve $2X^2 - X + 1 > 0$.

Solution: Let Y1 = represent the left side of the inequality and Y2 = the right side. We want to know <u>where</u> Y1 > Y2. Enter $2X^2 - X + 1$ at the Y1 = prompt and 0 at the Y2 = prompt. Press **[GRAPH]**. Since   Y2 = 0 is the X-axis, we do not "see" it as a separate line. It is, however, graphed. Your display should correspond to the display at the right.

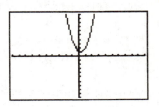

If the axes are turned "off" then the display clearly shows that the X-axis and Y2 = 0 are the same line. To turn off the axes, access the **FORMAT** menu. Press **[2nd]** **<FORMAT>** and cursor down and right until **AxesOff** is highlighted. Press **[ENTER]** and then **[GRAPH]**.

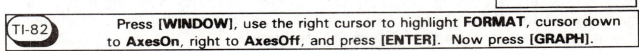

TI-82   Press **[WINDOW]**, use the right cursor to highlight **FORMAT**, cursor down to **AxesOn**, right to **AxesOff**, and press **[ENTER]**. Now press **[GRAPH]**.

TI-85/86   TO ACCESS THE **FORMAT** MENU, PRESS **[GRAPH]** **[MORE]** **[F3](FORMAT)** AND CURSOR DOWN AND RIGHT TO HIGHLIGHT **AxesOff**. PRESS **[ENTER]** TO CHOOSE THIS OPTION. PRESS **[F5(GRAPH)]**.

From this point on, make a mental note that Y1 is always being compared to the X-axis and do not enter Y2 = 0. **Think** about where Y1 is greater than Y2 (i.e. the X-axis). It is greater than the X-axis where it is **above** the axis. Use your highlighter pen to highlight the portion(s) of Y1 that are above the X-axis. Since all portions of Y1 are greater than the X-axis and since any real number is an acceptable value for X, the solution set is $\mathbb{R}$ (the set of real numbers). ◆

     Before proceeding, turn the axes back on.

Example 2: Solve the inequality $2X^2 - X + 1 < 0$ graphically.

Solution: Examine the graph displayed in Example 1. Where is $2X^2 - X + 1 < 0$ (i.e for what values of X is the graph below the X-axis)? Since the graph does not dip below the X-axis, there are no X-values for which $2X^2 - X + 1 < 0$. The solution is the empty set, ϕ or { }. ◆

Example 3: Solve $X^2 + 4X + 4 \leq 0$ graphically.

Solution: The algebraic statement indicates that the trinomial is less than **OR** equal to zero. Remember, 0 is represented by the X-axis. Enter $X^2 + 4X + 4$ after Y1 and sketch the graph that is displayed. **TRACE** along the path of the curve.

a. At what point(s) is the graph of $X^2 + 4X + 4$ **LESS THAN** 0? While tracing, determine the X-values for which the corresponding Y-values are negative. Clearly, there are none.

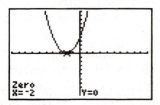

b. Use the ROOT/ZERO option to determine where the graph is **EQUAL TO** zero. Although this is an inequality, there is only **ONE** valid solution: X = -2.

◆

Consider the inequality $X^2 + 4X + 4 \geq 0$. You should be able to interpret the graphical display to determine that the solution set is all real numbers.
Next, consider the inequality $X^2 + 4X + 4 > 0$. Again, correct interpretation of the graphical display confirms the solution of all real numbers except for negative two.

GENERAL INEQUALITIES

In Example 3, $X^2 + 4X + 4 \leq 0$, the critical point in the solution set was X = -2. At X = -2, $X^2 + 4X + 4 = 0$. Negative two is a root of the equation. When solving inequalities, the critical points (i.e. the roots of the corresponding equation) will be the endpoints of the interval(s) of the solution region(s). The following example illustrates the procedure for solving quadratic inequalities. The way the graphical display is <u>interpreted</u> determines the solution to a given equation or inequality.

The steps for finding the solution set of an inequality will be demonstrated in the next example.

Example 4: Specify the solution set and the number line graph of the solutions of the inequality $X^2 - X - 6 < 0$.

Solution:
a. Be sure that the right hand side of the inequality is a ZERO.
b. Enter the polynomial $X^2 - X - 6$ at the Y1= prompt.
c. On the display at the right sketch the graphical representation of the solution.
d. Circle the X-intercepts (i.e. zeroes or roots) of the graph. These are the critical points. The circles will remain OPEN because the strict inequality symbol, "<", indicates that the critical points are not to be included as part of the solution set.

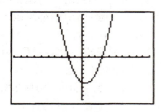

e. Use the ROOT/ZERO option (under the **CALC** menu) to determine the values of the X-intercepts. Label the values of these two critical points on the display.

f. Use your highlighter to highlight the section of the graph of $X^2 - X - 6$ that is **LESS THAN** 0 (i.e. below the X-axis).
g. The solution set is the set of all X values that yield the highlighted section of the graph. That is, $\{X \mid -2 < X < 3\}$.

Number line graph:

Alternate Option for TI-82/83/83plus/86 Users: The TABLE feature of the calculator is helpful in determining solution regions once the critical points have been calculated.

Set the table to start at a value of -2 (-2 was chosen because it is the critical point furthest to the left) with increments of 1. Press **[2nd]** **<TblSet>** to access this menu. Access the table by pressing **[2nd]** **<TABLE>**.

It is clear that Y1<0 when X has values larger than -2 and smaller than 3. Scrolling to X values smaller than -2 and larger than 3 confirms Y1 values that are positive (Y1 > 0) and are **not** solutions to the inequality.

We are concerned with solutions to the inequality $X^2 - X - 6 < 0$ and have entered $X^2 - X - 6$ at the Y1= prompt. We now know that when X < -2, Y1 > 0 and when X > 0, Y1 > 0. Therefore, these values should be shaded on the number line graph.

Use the steps outlined in Example 4 (or the TABLE) to solve $X^2 - X - 6 \geq 0$. Sketch the graph display, label the critical points, highlight the solution region(s), and interpret the solution.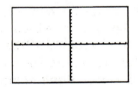

The interpretation of the graphical display expressed in set notation is $\{X \mid X \leq -2 \text{ or } X \geq 3\}$ and expressed as a number line graph is

EXERCISE SET

Use the steps outlined in a-g previously to graphically solve each of the following inequalities. For each problem you **must** sketch the graphical display and label the critical points. Record your solution in the following forms:
a. solution set   b. number line graph   c. interval notation

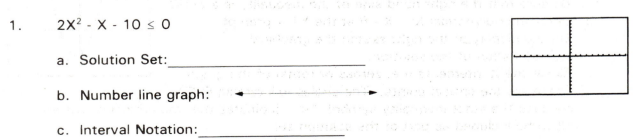

1. $2X^2 - X - 10 \leq 0$

a. Solution Set:_____

b. Number line graph:

c. Interval Notation:_____

2. $3X^2 + X - 4 \geq 0$ (Record critical points as fractions.)

   a. Solution Set:_____

   b. Number line graph: ←————————→

   c. Interval Notation:_____

3. $3X^2 + X - 4 < 0$ (Record critical points as fractions)

   a. Solution Set:_____

   b. Number line graph: ←———— ———→

   c. Interval Notation:_____

4. $X^2 + 10X + 25 \geq 0$

   a. Solution Set:_____

   b. Number line graph: ←————————→

   c. Interval Notation:_____

5. $X^2 + 10X + 25 > 0$

   a. Solution Set:_____

   b. Number line graph: ←————————→

   c. Interval Notation:_____

6. $4X^2 - 12X < -9$

   Solution Set:_____

7. $4X^2 - 12X + 9 > 0$

   a. Solution Set:_____

   b. Number line graph: ←———— ——→

   c. Interval Notation:_____

NOTE: Is the critical point actually 1.4999998 or 1.5? Attempting to convert to a fraction would seem to indicate that it is not 1.5. However, since the critical point(s) is the solution to the equation $4X^2-12X+9=0$, then the critical point should be the X value in the TABLE where Y1 = Y2. This can be accomplished in three ways:

a. Set the TABLE to start at X = 1.5 (the table increment is not critical since only one value is being checked). We want to know if Y1 = Y2 when X = 1.5. Press **[2nd]** <**TblSet**>, make this adjustment and then press **[2nd]** <**TABLE**>. At X = 1.5, Y1 = Y2. Thus X = 1.5 is the exact critical point; X = 1.4999998 is an approximation.

b. Use VALUE (**EVAL X**) to determine if Y1 = 0 when X = 1.5

When using the graph screen to solve equations/inequalities, you should be aware that the display coordinate values approximate the actual mathematical coordinates. The accuracy of these display values is determined by the height and width of the pixel space being displayed. The space height/width formulas are discussed in detail in the unit entitled "Preparing to Graph: Calculator Viewing Windows.

✍8. Explain the similarities and differences between graphically solving equations and inequalities.

✍9 Explain the significance of the critical points.

95

Solutions: 1. $\{X \mid -2 \leq X \leq 2.5\}$, 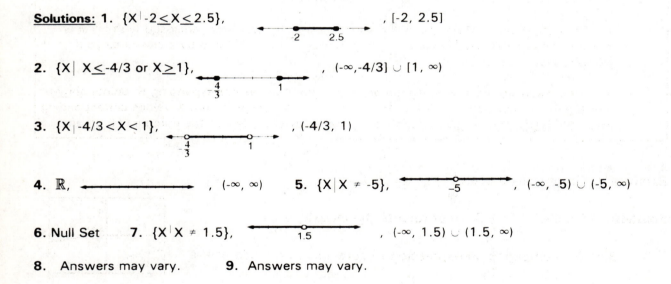 , $[-2, 2.5]$

2. $\{X \mid X \leq -4/3 \text{ or } X \geq 1\}$, , $(-\infty, -4/3] \cup [1, \infty)$

3. $\{X \mid -4/3 < X < 1\}$, , $(-4/3, 1)$

4. $\mathbb{R}$, , $(-\infty, \infty)$ **5.** $\{X \mid X \neq -5\}$, , $(-\infty, -5) \cup (-5, \infty)$

6. Null Set   **7.** $\{X \mid X \neq 1.5\}$, , $(-\infty, 1.5) \cup (1.5, \infty)$

8. Answers may vary.   **9.** Answers may vary.

96

UNIT 15
GRAPHICAL SOLUTIONS: RATIONAL INEQUALITIES

* Unit 14 is a prerequisite for the unit. Answers appear at the end of the unit.

The previous unit examined solving quadratic inequalities. To solve <u>rational</u> nonlinear inequalities we will use some of the same procedures from the previous unit and investigate necessary modifications.

The following steps were used in solving quadratic inequalities.

SOLVING QUADRATIC INEQUALITIES GRAPHICALLY

a. The inequality was written with one side (usually the left) greater than/less than 0.

b. The left side of the inequality was entered on the calculator at the Y1 = prompt. The standard viewing screen was used to view the graph.

c. Because the right side of the inequality was equal to zero, the X-axis was used as a reference.

d. Once the graph was displayed, the critical points were found by using the ROOT/ZERO option under the **CALC** menu. Recall these were the values of X that made the equation associated with the given inequality a true statement. Graphically, they were the point(s) where the graph crossed the X-axis, the x-intercepts.

e. The critical points were graphed on a number line. They were enclosed with an open circle if the original inequality was a <u>strict</u> inequality (> or <) and by a closed circle if equality was included (≥ or ≤).

f. If the inequality was >, the solutions were the X-values corresponding to points <u>above</u> the X-axis. Conversely, if the inequality was <, solutions were the X-values corresponding to points <u>below</u> the X-axis. Appropriate regions were shaded on the number line graph.

Example 1: Graphically solve $\dfrac{12}{X} \geq 3$.

Solution: To solve $\dfrac{12}{X} \geq 3$, first rewrite the inequality as

$\dfrac{12}{X} - 3 \geq 0$ (comparing an expression to zero allows the use of the

X-axis as a reference point.) Enter $\dfrac{12}{X} - 3$ at the Y1 = prompt and

view the graph in the standard viewing window. Your display should match the one above.

Find the critical point(s). First, circle the point where the graph crosses the X-axis on the display. Now use the ROOT/ZERO option of the calculator to find the critical point.

To confirm that 4 is a root (critical point), access the TABLE (first set the table to begin at - 1 and increment by 1). When X = 4, Y1 has a value of 0. This confirms that 4 is the root (and thus a critical point). However, notice that when X = 0 that Y1 = ERROR. There is an ERROR message for this value of X. Because the fraction $\frac{12}{X}$ is undefined when X = 0, this value for X IS NOT part of the solution. This is an excluded (or restricted) value.

a. Locate the excluded value (0) and the root (4) on the graph.

Notice that the coordinate 0 is marked with an open circle, while the coordinate 4 is marked with a closed circle. These are different because 4 is a solution to the inequality, whereas the value 0 cannot be a solution to the inequality because it results in an expression that is undefined.

IT IS IMPERATIVE THAT ALL EXCLUDED VALUES (RESTRICTED VALUES) ARE DETERMINED **BEFORE** GRAPHING. Recall, these are the values that make any denominator equal 0. They must be located on the number line and are ALWAYS marked by an open circle as they are NEVER a part of the solution.

b. Access the graph and **shade** the part(s) of the pictured number line that correspond to those points of the graph that are above the X-axis. Shade the corresponding X-values on the number line below.

Your number line graph should correspond to {X | 0 < X ≤ 4} which is expressed as (0,4] in interval notation.

c. Accessing the TABLE feature allows us to check the solution. Notice that for values of X SMALLER than 0, the Y1 values are negative, and are not a part of the desired solution. When X is GREATER than 0 but LESS than 4 the corresponding Y1 values are positive, and thus are solutions to $\frac{12}{X} - 3 \geq 0$.

Scrolling past 4 (to X values GREATER than 4) the corresponding Y1 values are negative, and are not a part of the desired solution.

Example 2: Graphically solve the inequality $\frac{-8}{X-4} \leq \frac{5}{4-X}$.

Solution: a. Rewrite the inequality so that one side equals zero: $\frac{-8}{X-4} - \frac{5}{4-X} \leq 0$

Enter the left side at the Y1 = prompt and press **[GRAPH]**. Your display should match the one at the right.

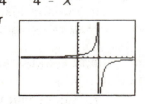

b. Locate the excluded value of 4 by setting denominator factors equal to 0 and solving for X.

c. Now find any critical point(s). These are the point(s) where the graph crosses the X-axis. Use the ROOT/ZERO option to **try** to find the root(s).

98

d. To better see what is going on, put the calculator in **DOT** mode. To do this, press **[MODE]** and cursor down to **Connected** and then right to **Dot**. Press **[ENTER]** to highlight this mode. Press **[GRAPH]**. Your display should correspond to the display at the right.

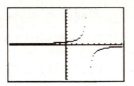

The TI-83/83plus/86 allow the personalization of graphs. Press **[Y=]** and use the arrow key to place the cursor over the graphing icon to the left of Y1. Press **[ENTER]** while observing the change in the graphing icon. When the diagonal pattern of "dots" slanting left to right is displayed, stop. Press **[GRAPH]** and compare your display to the one pictured. If you initially changed **MODE** to DOT then the calculator automatically set the graphing icon to **dot**. TI-83/83plus and TI-86 users should be aware that the graphing icon will override the CONNECTED/DOT **MODE**. Thus, if the **MODE** screen is set to DOT but the graphing icon is set to **solid** then all points will be connected when graphed.

TI-85	PRESS **[GRAPH]** **[MORE]** **[F3](FORMAT)**. CURSOR DOWN AND OVER TO HIGHLIGHT **DRAWDOT** AND PRESS **[ENTER]**.

TI-86	THIS CALCULATOR ALLOWS THE PERSONALIZATION OF GRAPHS. BE SURE THE GRAPH MENU IS DISPLAYED AT THE BOTTOM OF YOUR SCREEN. PRESS **[F1](Y(X)=)** **[MORE]** **[F3](STYLE)**. CONTINUE PRESSING (STYLE) UNTIL THE DIAGONAL PATTERN OF "DOTS" SLANTING LEFT TO RIGHT IS DISPLAYED. PRESS **[M5](GRAPH)** AND COMPARE YOUR DISPLAY TO THE ONE PICTURED. IF YOU INITIALLY CHANGED **MODE** TO DOT THEN THE CALCULATOR AUTOMATICALLY SET THE GRAPHING ICON TO DOT. TI-86 USERS SHOULD BE AWARE THAT THE GRAPHING ICON WILL OVERRIDE THE CONNECTED/DOT **MODE**. THUS, IF THE **MODE** SCREEN IS SET TO **DOT** BUT THE GRAPHING ICON IS SET TO **SOLID** THEN ALL POINTS WILL BE CONNECTED WHEN GRAPHED.

When comparing the two graphical displays (connected mode vs. dot), notice the vertical line (where we believed there was a root) is gone. In CONNECTED MODE this line connected two adjacent pixel points on the graph. However, in DOT MODE it is clear that these two points are at opposite ends of the graph and should not be connected. To connect them would mean that 4 is an acceptable value for X.

The graph <u>never</u> crosses the X-axis but rather jumps from a location above the X-axis to one below it. **TRACE** and observe the X-values to confirm this.

e. Recall, we want the solutions to the inequality $\dfrac{-8}{X-4} - \dfrac{5}{4-X} \le 0$. Place 4 on the number line and circle it (an excluded value) and shade to the right:

If you are unsure as to why the solution shades to the right, use the VALUE (Eval) feature of your calculator to test points to the right and left of the critical point of 4.

The solution set is $\{X \mid X > 4\}$ or $(4, \infty)$ in interval notation.

f. Accessing the TABLE feature confirms the fact that values of Y1 are positive for X less than 4 and negative for X greater than 4, thus validating the solution above. ◆

The following steps outline the method of finding the solution(s) to the inequality .

SOLVING RATIONAL INEQUALITIES GRAPHICALLY

a. Rewrite the inequality with one side equal to 0. Enter the non-zero side at the Y1 = prompt. Remember, the calculator should be in DOT MODE or the dot style icon should be activated.

b. Find the excluded values by setting each denominator equal to 0 and solving for the X values.

Enter these values on the number line and mark them with <u>open circles</u> so that they are not inadvertently included in the solution.

c. Find the roots/zeroes of the equation associated with the inequality by using the ROOT/ZERO option under the **CALC** menu OR by solving the equation associated with the inequality, i.e. change the inequality symbol to an equal sign.

Enter the roots on the number line and mark them with closed circles if the inequality is $\leq$ or $\geq$, and with open circles if the inequality is $<$ or $>$.

d. Shade the regions on the number line that represent the x-values of the ordered pairs on the graph where the y-values are greater than/less than 0 as determined by the inequality. You may also determine the appropriate x-values by testing values in each region of the number line. You can accomplish the same thing by using the TABLE.

e. Write the solution as a solution set and in interval notation.

f. Reset the calculator to CONNECTED MODE and compare the number line graph to the graphical display on the calculator screen. **TRACE** and observe X and Y values. Confirm that the solution above is correct. Notice that when TRACING left the Y values jump from negative values to positive values as the screen scrolls.

EXERCISE SET

Directions: Use the steps outlined above to solve each inequality below. Begin by setting the calculator in Connected MODE with a standard viewing WINDOW. Remember to access the TABLE feature (when appropriate) and convert to dot mode as **YOU** deem appropriate. Use the combination of algebra and calculator that makes you feel comfortable.

1. $$\frac{X^2 + 6X + 9}{X + 5} > 0$$

   a. Number line graph: ← —————————— →

   b. Solution set: _____

   c. Interval notation: _____

2. $\dfrac{2}{X-2} < \dfrac{3}{X}$

    a. Number line graph: ◄ ─────────── ►

    b. Solution set: _____

    c. Interval notation: _____

3. $\dfrac{(2X-1)(X-5)}{X+3} \le 0$

    a. Number line graph: ◄ ─────────── ►

    b. Solution set: _____

    c. Interval notation: _____

Note: In #3, did you shade _only_ between the roots of 1/2 and 5? This problem illustrates an important point: ALWAYS look to the left (or right) of the restricted values and/or critical points to see how the graph behaves.

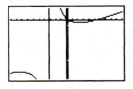

Set the calculator in Connected MODE. TRACE left on the graph of the above expression and go beyond the vertical line that connects the two non-adjacent pixel points. You will not see any more graph (TRACE until your graph shifts <u>at least</u> once). Look at the X and Y coordinates at the bottom of the screen, and then check the window values. We will now scroll down to Ymin and enter -50. Press **[GRAPH]** and compare your display to the one pictured.

NOW you should understand why the region of the number line less than -3 is shaded.

✍4. Explain what excluded values are, how to find them, and their inclusion or non-inclusion in solutions.

✍5. Discuss the advantages and disadvantages of looking at solutions of non-linear rational inequalities in both Dot and Connected MODE.

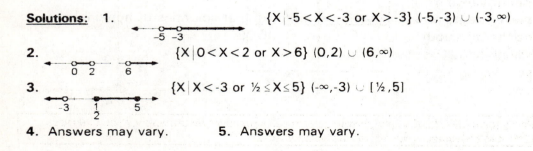

Solutions: **1.** $\{X \mid -5 < X < -3 \text{ or } X > -3\}$ $(-5,-3) \cup (-3,\infty)$

2. $\{X \mid 0 < X < 2 \text{ or } X > 6\}$ $(0,2) \cup (6,\infty)$

3. $\{X \mid X < -3 \text{ or } \frac{1}{2} \le X \le 5\}$ $(-\infty,-3) \cup [\frac{1}{2},5]$

4. Answers may vary.   **5.** Answers may vary.

UNIT 16
HOW DOES THE CALCULATOR ACTUALLY GRAPH?
(EXPLORING POINTS AND PIXELS)

*Unit 3 is a prerequisite for this unit.

Consider the first degree polynomial expression $\frac{3}{4}$X + 6. Use the STOre feature of your

graphing calculator to evaluate the polynomial for each of the given values of X:

X	-8	-4	0	4	8
(3/4)X + 6					

The STOre feature is used to evaluate polynomials for given values of the variable(s). The
y = edit screen can quickly accomplish the same task if the polynomial is a one variable
polynomial. For calculators with no table, the STOre feature allows the user to generate a
table of values. TI-82/83/83plus/86 calculators have a TABLE feature that allows the user
to store a polynomial in one variable, and evaluate that polynomial for
multiple values. Press **[Y =]** and enter (store) the expression
(3/4)X + 6 after the Y1 = prompt. Press **[2nd] <TblSet>** and set
the table to start at -8 to correspond with the chart constructed
above. You may set the table increment (ΔTbl) to 4 or you may
leave it at 1 and simply cursor down through the table. Pressing **[2nd]**
<TABLE> will now display the table at the right.

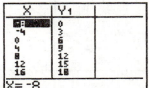

The expression stored at Y1 has been evaluated for each value of the variable *X* thus

creating ordered pair solutions (x,y) to the equation Y = $\frac{3}{4}$X + 6. Plot each of the ordered

pairs from chart/table above on the graph at the right. Use a
ruler to connect the points to graph the line of the equation
Y = (3/4)X + 6.
Compare the equation to the calculator graph by pressing
[ZOOM] [6:ZStandard]. This displays
the graph in the standard viewing
window. The standard viewing
window sets the maximum value of
each axis at 10 and the minimum
value at -10. Pressing **[WINDOW]**
will allow you to see the values of
the viewing window.

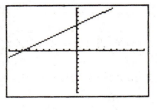

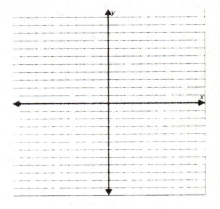

Since the calculator drawn graph is not a *straight* line as expected, we will construct a graph in the same manner as the calculator to understand what is happening.

Pretend the graph at the right is the calculator screen and that each box represents a pixel space. "Light up" the X-axis and Y-axis by <u>lightly</u> shading a horizontal line of boxes and a vertical line of boxes where the two axes should be located.

Now *plot* the X-intercept of (-8,0) and the Y-intercept of (0,6) by <u>darkly</u> shading the appropriate pixel space (box).

Place your ruler so that it connects these two pixels. Now <u>darkly</u> shade in a path of pixels along the edge of your ruler. Remember, the entire square representing the pixel must be shaded.

The graphs below show the comparison between your hand drawn graph using the ordered pairs, the actual calculator display and your pixel sketch.

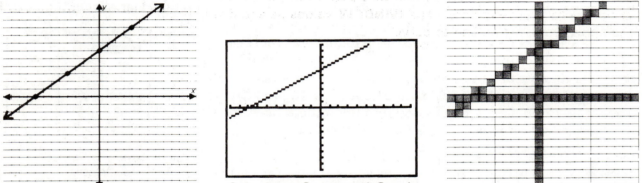

| Hand Drawn Graph | Calculator Generated Graph | Pixel Sketch |

The calculator plots points by lighting up little squares on the screen called pixels. The TI-82/83/83plus screen is 95 pixel points wide (with 94 spaces between the horizontal pixel points) by 63 pixel points high (with 62 spaces between the vertical pixel points). Because there are only a finite number of pixel spaces to light up, the calculator may only be able to "light up" a pixel that is <u>close</u> to the desired point.

| TI-85/86 | THE TI-85/86 SCREEN IS 127 PIXEL POINTS WIDE (WITH 126 SPACES BETWEEN HORIZONTAL PIXEL POINTS) BY 63 PIXEL POINTS HIGH AS IN THE TI-82/83 SCREEN. |

Now that you have seen how the calculator must light up pixels to graph a straight line, we will examine what happens to curves.

A semi-circle with a radius of 5 units has been drawn on the graph at the right. Shade in the squares along the path of the semi-circle to simulate the calculator "lighting up" pixel spaces. The pixels representing the X and Y axes have already been shaded for you.

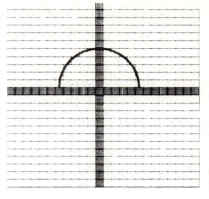

Now graph a semicircle on the calculator by entering the expression $\sqrt{25 - X^2}$ on the **y =** screen. Press [Y =] and be sure you enter the expression as $\sqrt{}$ **(25 - X²)** at the Y1 = prompt. Press [**GRAPH**].

TI-85/86	Press [GRAPH] [F1](y(x) =) to enter $\sqrt{}$ **(25 - X²)** and to display the graph, press [2ND] <M5> (GRAPH).

Your graph should look like the one displayed at the right.

Notice how flat the top of the semi-circle appears. All calculator drawn lines and curves will consist of a pattern of boxes, and vertical or horizontal line segments. The **WINDOW** values selected will affect the appearance of the line or curve.

UNIT 17
PREPARING TO GRAPH: CALCULATOR VIEWING WINDOWS

*Unit 16 is a prerequisite for this unit. Answers appear at the end of the unit.

TI-85/86	IF USING THE TI-85/86, GO TO THE GUIDELINES (PG.115).

SETTING UP THE GRAPH DISPLAY

The calculator's display is controlled through the **MODE** and **FORMAT** screens.

Press the [**MODE**] key. **MODE** controls how numbers and graphs are displayed and interpreted. The current settings on each row should be highlighted as displayed. The blinking rectangle can be moved using the 4 **cursor** (arrow) keys. To change the setting on a particular row, move the blinking rectangle to the desired setting and press [**ENTER**].

NOTE: Items must be highlighted to be activated.

Normal vs. Scientific notation
Floating decimal vs. Fixed to 9 places
Type of angle measurement
Type of graphing: function, parametric, polar, sequence
Graphed points connected or dotted
Functions graphed one by one
Numbers can be viewed as Real or complex
Screen can be split to view two screens simultaneously

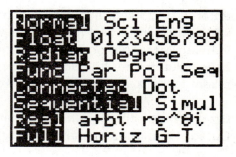

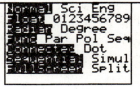

(TI-82) The first six options are the same as the TI-83/83plus. The TI-82 will not compute with **complex numbers.**

To return to the home screen press [**CLEAR**] or [2nd] <QUIT>.

Press [**Y=**]. The calculator can graph up to 10 different equations at the same time. Because **MODE** is in the sequential setting, the graphs will be displayed sequentially. Note that cursoring down accesses additional Y= prompts. The display of the equations entered on the **Y=** screen is controlled by the size of the viewing window. The dimensions of the viewing window are determined by the values entered on the **WINDOW** screen.

The TI-83/83plus/86 calculators have a feature called the *graph style icon* that allows you to distinguish between the graphs of equations. To change styles (on the TI-83/83plus), press [**Y=**] and observe the "\" in front of each Y. Use the left arrow to cursor over to this "\." The diagonal should now be moving up and down. Pressing **[ENTER]** once changes the diagonal from thin to thick. The graph of Y1 will now be displayed as a thick line. Repeatedly pressing **[ENTER]** displays the following:

- ◥: shades above Y1
- ◣: shades below Y1
- -o: traces the leading edge of the graph followed by the graph
- o: traces the path but does not plot
- ⋅⋅: displays graph in dot, not connected MODE

     It is important to remember that the graphing icon takes precedence over the **MODE** screen. If the icon is set for a solid line and the **MODE** screen is set for DOT and not solid, the graphing icon will determine how the graph is displayed. Pressing **[CLEAR]** to delete an entry at the Y = prompt will automatically reset the graphing icon to default, a solid line.

Press **[ZOOM] [6:ZStandard] [WINDOW]**. This is called the standard viewing window. The information on this screen indicates that in a rectangular coordinate system the X-values will range from -10 to 10 and the Y-values will range from -10 to 10. The interval notation for this is [-10,10] by [-10,10]. The Xscl = 1 and Yscl = 1 settings indicate that the tic marks on the axes are one unit apart. The values

```
WINDOW
 Xmin=-10
 Xmax=10
 Xscl=1
 Ymin=-10
 Ymax=10
 Yscl=1
 Xres=1
```

entered on this screen may be changed by using the cursor arrows to move to the desired line and typing over the existing entry. When drawing a graph, you may set the desired viewing rectangle on the calculator as well as scale the X-axis and Y-axis. The row labeled Xres = determines the screen resolution. It should be set equal to 1, which means that each pixel on the X-axis will be evaluated and graphed.

Graphs on a rectangular coordinate system
Cursor location is displayed on screen
Graphing grid is not displayed
Axes are visible
Axes are not labeled with an X and Y
The expression entered at Y1 is displayed on the graph when trace is activated.

TI-82 Be sure the cursor is on the word **WINDOW** and use the right arrow key to cursor over to **FORMAT**. The following settings should be highlighted:

Graphs on a rectangular coordinate system
Cursor location is displayed on screen
Graphing grid is not displayed
Axes are visible
Axes are not labeled with an X and Y

EXERCISE SET

Directions: Before proceeding further, press [Y =] and clear all entries.

1. Press **[WINDOW]** and enter Xmin = -5, Xmax = 5, Xscl = 1, Ymin = -12, Ymax = 7, Yscl = 1. Be sure to use the gray **[(-)]** key for negative signs. Press **[GRAPH]** to view the coordinate axes. Count the tic marks on the axes and see how these marks correspond to the max and min values. Label the last tic mark on each axis (i.e. farthest tic mark left, right, up and down) with the appropriate integral value.

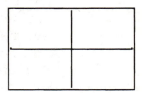

| TI-85/86 | TI-85/86 USERS MUST PRESS **[CLEAR]** TO DELETE MENU DISPLAY BEFORE COUNTING TIC MARKS. |

2. Change the viewing window to Xmin = -20, Xmax = 70, Xscl = 10, Ymin = -5, Ymax = 15, Yscl = 3 and press **[GRAPH]** to view the axes. How many tic marks are on the positive portion of the X-axis?_____ How many units does each of the tic marks on this axis represent?_____ Based on your last two answers, how many units long is the positive portion of the X-axis?_____ Does this number correspond to the Xmax value given in the problem? _____

   In your own words, explain what is happening.

3. To help "de-bug" errors in graphing set ups later on, describe what you think would happen if Xmin = 10 and Xmax = -5. You might want to draw your own set of coordinate axes and <u>try</u> to label them in this manner. Enter these values on the **WINDOW** screen and press **[GRAPH]**. What did happen?

4. Reset Xmin = -10, Xmax = 10 and describe what you think will happen if you set Ymin = 5, Ymax = 5. Again, enter the values and press **[GRAPH]**. What did happen? (Try drawing your own set of axes and labeling them as indicated.)

5. What should the relationship between Max and Min be? (i.e. Min > Max, Min < Max, or Min = Max)

ENTERING EXPRESSIONS TO BE GRAPHED: THE [Y=] KEY

Reset the viewing window to ZStandard, by pressing **[ZOOM]** **[6:ZStandard]**.

TI-85/86   TI-85/86 USERS PRESS **[GRAPH]** **[F3](ZOOM)** **[F4](ZSTD)**.

Press **[Y=]**. On the screen Y1= is followed by a blinking cursor. Anything else can be cleared by pressing **[CLEAR]**. Enter the expression -2X+ 6 at the Y1= prompt and press **[ENTER]**. The cursor is now on the second line following Y2=. At this prompt, enter the expression (1/2)X - 4. Note that the equal signs beside both Y1 and Y2 are highlighted. This means that both equations will be graphed. Press **[GRAPH]** to display the graph screen. See display.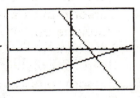

Note: Y = (1/2)X - 4 would be displayed in your text as $y = \dfrac{1}{2}x - 4$.

GRAPHING

To graph Y = -2X + 6 only, press **[Y=]** and use the arrow key to move the cursor over the equal sign beside Y2. Press **[ENTER]**. Notice that the equal sign beside Y2 is *not* highlighted, whereas the equal sign beside Y1 *is* highlighted. Press **[GRAPH]**; only the highlighted equation, Y1, is graphed.

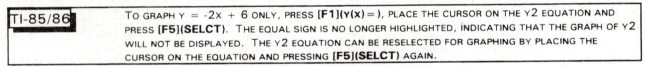

TI-85/86   TO GRAPH Y = -2X + 6 ONLY, PRESS **[F1](y(x)=)**, PLACE THE CURSOR ON THE Y2 EQUATION AND PRESS **[F5](SELCT)**. THE EQUAL SIGN IS NO LONGER HIGHLIGHTED, INDICATING THAT THE GRAPH OF Y2 WILL NOT BE DISPLAYED. THE Y2 EQUATION CAN BE RESELECTED FOR GRAPHING BY PLACING THE CURSOR ON THE EQUATION AND PRESSING **[F5](SELCT)** AGAIN.

On the viewing screen at right, the calculator draws a set of axes whose minimum and maximum values and scale match the choices under **WINDOW**. The graph of Y1 is drawn from left to right. Return to the Y= menu by pressing **[Y=]**. Cursor down to Y2 and *turn on* this graph by highlighting the equal sign. Press **[GRAPH]** and notice that the two graphs are drawn in sequence. **SEQUENTIAL** was chosen from the **MODE** menu earlier.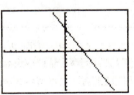

TI-85/86   TI-85/86 USERS GO TO THE GUIDELINES (PG.116), AND READ THE SECTION ENTITLED ALTERING THE VIEWING WINDOW.

ALTERING THE VIEWING WINDOW

The last unit addressed the size of the calculator screen (viewing window). Because the screen is 95 pixel points wide by 63 pixel points high, there are 94 horizontal spaces and 62 vertical spaces to light up. When tracing on a graph, the readout changes according to the size of the space. The size of the space can be controlled by the following formulas:

$$\frac{Xmax - Xmin}{94} = \text{horizontal space width}, \quad \frac{Ymax - Ymin}{62} = \text{vertical space height}.$$

We will examine some preset viewing windows and how they affect the pixel space size.

Press **[ZOOM]**. There are ten entries on this screen. (The TI-82 has nine entries.) The down arrow key can be used to view remaining entries.

1: Boxes in and enlarges a designated area.
2: Acts like a telephoto lens and "zooms in."
3: Acts like a wide-angle lens and "zooms out."
4: Cursor moves are ONE tenth of a unit per move.
5: "Squares up" the previously used viewing window.
6: Sets axes to [-10,10] by [-10,10].
7: Used for graphing trigonometric functions.
8: Cursor moves are ONE integer unit per move.
9: Used when graphing statistics.
0: Replots function, recalculating Ymin and Ymax.

ZDecimal is useful for graphs that require the use of the calculator's TRACE feature. Applying the horizontal space width formula,

$$\frac{Xmax - Xmin}{94} = \frac{4.7 - (-4.7)}{94} = 0.1,$$ changes the X-values by one-

tenth of a unit each time the cursor is moved. This is why this screen yields "friendly" values when TRACING. (In general, Xmax - Xmin needs to be a multiple of 94 to produce a "friendly" screen.)

TI-85/86 TI-85/86 USERS ARE REMINDED THAT THE DENOMINATOR OF THE HORIZONTAL SPACE WIDTH FORMULA SHOULD BE 126. IN GENERAL, xMAX - xMIN NEEDS TO BE A MULTIPLE OF 126 TO PRODUCE A "FRIENDLY" SCREEN.

Enter Y1 = -2X + 3 and Y2 = ½X - 2 on the Y = screen. Press **[ZOOM] [4:ZDecimal]** (TI-85/86 users press **[GRAPH] [F3] [MORE] [F4] (ZDECM)**)to display the graph of these two lines in the ZDecimal viewing window. TRACE along the graph of one of the lines and observe the changes in X values. The change should be one-tenth of a unit. (Remember, the Y-values are dependent on the values selected for X.)

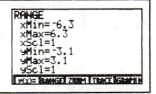

ZInteger is useful for application problems where the X-value is valid only if represented as an integer (such as when X equals the number of tickets sold, number of passengers in a

vehicle, etc.). The horizontal space width formula, $\frac{Xmax - Xmin}{94} = \frac{47 - (-47)}{94} = 1$, changes

the X-values by one unit each time the cursor is moved.

TI-85/86 REMEMBER, THE TI-85/86 SCREEN IS 127 PIXELS WIDE THUS THE HORIZONTAL SPACE WIDTH FORMULA STILL NEEDS A DIVISOR OF 126.

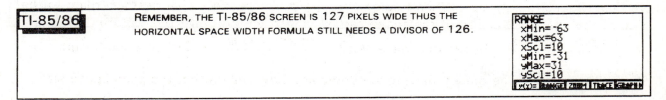

Press **[ZOOM] [8:ZInteger]** , move the cursor to the origin of the graph (x = 0 and y = 0), and press **[ENTER]** to display the graph of these two lines as indicated at the right.

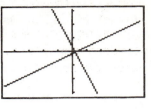

TI-85/86   PRESS **[GRAPH] [ZOOM] [MORE] [MORE]** **(ZINT)** AND MOVE THE CURSOR AS DIRECTED ABOVE.

TRACE along the graph of one of the lines and observe the changes in X values. The change should be one unit.

ZStandard provides a good visual comparison between hand sketched graphs (or textbook graphs) that are approximately [-10,10] by [-10,10]. Applying the horizontal space width formula,

$$\frac{Xmax - Xmin}{94} = \frac{10 - (-10)}{94} \approx 0.212765974,$$ the X-values will

```
WINDOW
  Xmin=-10
  Xmax=10
  Xscl=1
  Ymin=-10
  Ymax=10
  Yscl=1
  Xres=1
```

change by .212765974 each time the TRACE cursor is moved. If you are using the TRACE feature, you will usually want a screen with "friendlier" X-values than this one provides.

Press **[ZOOM] [6:ZStandard]** to display the graph of these two lines as indicated at the right. TRACE along the graph of one of the lines and observe the changes in x values. If you select two consecutive x-values and find the difference, it should be 0.212765974.

ZDecimal x n: ZDecimal frequently does not provide a large enough viewing window. When this is the case, you may multiply the Xmin and Xmax by the same constant and the Ymin and Ymax by the same constant to produce a larger viewing rectangle which still provides cursor moves in tenths of units. Multiplying Xmin and Xmax by 2 would mean cursor moves of two-tenths of a unit:

```
WINDOW
  Xmin=-9.4
  Xmax=9.4
  Xscl=1
  Ymin=-6.2
  Ymax=6.2
  Yscl=1
  Xres=1
```

$$\frac{Xmax - Xmin}{94} = \frac{2(4.7) - 2(-4.7)}{94} = 0.2,$$ whereas multiplying by 3 would mean cursor

moves of three-tenths of a unit. The screen above is the ZDecimal screen with the max and min values multiplied by 2. This WINDOW will be referred to in the future as ZDecimal x 2.

TI-85/86   REMEMBER, THE TI-85/86 SCREEN IS 127 PIXELS WIDE THUS THE HORIZONTAL SPACE WIDTH FORMULA STILL NEEDS A DIVISOR OF 126.

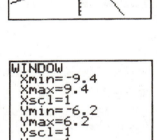

Press **[Y =]** and clear all entries. Enter Y1 = $\frac{2}{3}$X + 6 and Y2 = $-\frac{3}{2}$X - 5 . (Did you remember to put parentheses around the fractions?)

$Y1 = \dfrac{2}{3}X + 6$ and $Y2 = -\dfrac{3}{2}X - 5$ are perpendicular lines whose

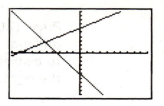

intersection forms a 90° angle. Press **[ZOOM] [6:ZStandard]** for the standard viewing rectangle. Notice that the lines do not appear to be perpendicular. This is because the screen is rectangular - not square.

| TI-85/86 | TI-85/86 USERS SHOULD PRESS **[GRAPH] [F3](ZOOM) [F4](ZSTD)** TO AUTOMATICALLY SET THE STANDARD VIEWING RECTANGLE. |

Press **[ZOOM]** again and this time select **[5:ZSquare]**. ZSquare *squares up* the viewing screen based on the previous viewing window. The lines should now appear to be perpendicular. Notice that the tic marks are all evenly spaced now.

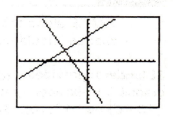

EXERCISE SET

 6.　Press **[WINDOW]** to see how the Max and Min values were affected by the Zsquare command applied above. Enter the WINDOW values displayed. Explain how the viewing window is different from the ZStandard viewing window.

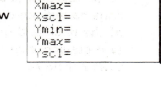

NOTE: The ZDecimal screen (or any multiplicity of this screen) will provide a "squared up" graph screen on the TI-82/83/83plus. TI-85/86 users will need to use ZSquare .

7.　**CLEAR** all entries on the **Y=** screen. Using

$Y1 = \dfrac{1}{2}X - 4$, press **[ZOOM] [4:ZDecimal]**

and at the right, sketch the screen as displayed. Write the appropriate value for the "endpoint" of each axis on the graph.

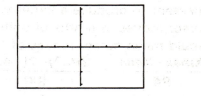

8.　Now, press **[ZOOM] [8:ZInteger]** (pause for your graph to be displayed) **[ENTER]** and at the right, sketch the screen as displayed. Write the appropriate value for the "endpoint" of each axis on the graph.

NOTE: ZDecimal and ZStandard do not require you to press [ENTER] to activate the viewing WINDOW, but ZInteger demands it.

✍9.   In your own words, explain the differences in the displays in #7 and #8. What accounts for these differences?

10.   Enter $12X^6 - 58X^4 + 84X^2 + 8$ at the **Y1=** prompt. Try to view the graph in each of the pre-set viewing windows discussed in this unit - ZDecimal, ZStandard, ZInteger, then sketch the graph as displayed in each of the indicated viewing WINDOWS. Both the Xscl and Yscl should be equal to zero.

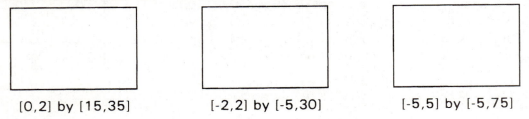

[0,2] by [15,35]   [-2,2] by [-5,30]   [-5,5] by [-5,75]

Pre-set viewing WINDOWS can provide a "starting point" for displaying a complete graph but frequently do not display all of the critical features of the graph. The next unit examines in detail the approach necessary for setting a good viewing WINDOW for individual graphs.

Solutions: **1.** left: -5, right: 5, top: 7, bottom: -12 **2** 7, 10, 70, yes; By increasing the value of the scale, the axes are increased in size without physically extending the length.

3. Because the maximum does not exceed the minimum (i.e. -5 is NOT greater than 10) the calculator displayed an error message. Moreover, that error message tells you that _you_ made an error in setting the values.

4. The y-axis has not been given a defined length by setting the max and min at the same value. Again, an error message is displayed.

5. Max > Min   **6.** To have a "square" screen, tic marks must be evenly spaced. This was accomplished by adding tic marks to the X-axis.

7. 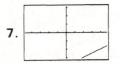   **8.**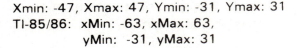

Xmin: -4.7, Xmax: 4.7, Ymin: -3.1, Ymax: 3.1   Xmin: -47, Xmax: 47, Ymin: -31, Ymax: 31
TI-85/86: xMin: -6.3, xMax: 6.3,   TI-85/86: xMin: -63, xMax: 63,
          yMin: -3.1, yMax: 3.1   yMin: -31, yMax: 31

9. The difference was the amount and position of graph displayed. More of the graph was displayed on the ZInteger screen. This was because the _scales_ were different on the two screens.

SETTING UP THE GRAPH DISPLAY

The calculator's display is controlled through the **MODE** and **GRAPH/FORMAT** screens.

Press **[2nd] <MODE>**. The current settings on each row should be highlighted as displayed. The blinking rectangle can be moved using the 4 **cursor** (arrow) keys. To change the setting on a particular row, move the blinking rectangle to the desired setting and press **[ENTER]**.

NOTE: Items must be highlighted to be activated.

Normal vs. Scientific notation
Floating decimal vs. Fixed to 11 places
Type of angle measurement
Complex number display
Type of graphing: function, polar, parametric, differential eq.
Performs computations in bases other than base 10
Format of vector display
Type of differentiation

The **FORMAT** screen is accessed by pressing **[GRAPH] [MORE] [F3](FORMT)**. The following settings should be highlighted.

Graphs on both rectangular and polar coordinate system
Cursor location is displayed on the screen
Graphed points are connected or discrete
Functions displayed sequentially or simultaneously
Graphing grid is not displayed
Axes are visible
Axes are not labeled with "x" and "y"

To return to the home screen, at this point, press **[CLEAR]** or **[EXIT]**.

Press **[GRAPH] [F1](y(x)=)**. The calculator can graph up to 99 different equations at the same time. Because **MODE** is in the sequential setting, the graphs will be displayed sequentially. The display of the equation(s) entered on the y(x)= screen is controlled by the size of the viewing window. The dimensions of the viewing window are determined by the values entered on **RANGE** screen, which is referred to as the **WINDOW** screen in the core units.

TI-86

THIS CALCULATOR HAS A FEATURE CALLED THE "GRAPH STYLE ICON" THAT ALLOWS YOU TO DISTINGUISH BETWEEN THE GRAPHS OF EQUATIONS. PRESS **[GRAPH] [F1](Y(X)=)** AND OBSERVE THE "\" IN FRONT OF EACH Y. PRESS **[MORE]** FOLLOWED BY **[F3](STYLE)**. THE DIAGONAL SHOULD NOW HAVE CHANGED FROM THIN TO THICK. THE GRAPH OF Y1 WILL NOW BE DISPLAYED AS A THICK LINE. REPEATEDLY PRESSING **[STYLE]** DISPLAYS THE FOLLOWING:

- ◥: SHADES ABOVE Y1
- ◣: SHADES BELOW Y1
- -O: TRACES THE LEADING EDGE OF THE GRAPH FOLLOWED BY THE GRAPH
- O: TRACES THE PATH BUT DOES NOT PLOT
- ∴: DISPLAYS GRAPH IN DOT, NOT CONNECTED MODE

IT IS IMPORTANT TO REMEMBER THAT THE GRAPHING ICON TAKES PRECEDENCE OVER THE **MODE** SCREEN. IF THE ICON IS SET FOR A SOLID LINE AND THE **MODE** SCREEN IS SET FOR DOT AND NOT SOLID, THE GRAPHING ICON WILL DETERMINE HOW THE GRAPH IS DISPLAYED. PRESSING **[CLEAR]** TO DELETE AN ENTRY AT THE Y = PROMPT WILL AUTOMATICALLY RESET THE GRAPHING ICON TO DEFAULT, A SOLID LINE.

Press **[GRAPH] [F3](ZOOM) [F4](ZSTD) [2nd] <M2>(RANGE)**. (WIND on the TI-86.) This is called the standard viewing window. The information on this screen indicates that in a rectangular coordinate system the x-values will range from -10 to 10 and the y-values will range from -10 to 10. The interval notation for this is [-10,10] by [-10,10]. The xScl = 1 and yScl = 1 settings indicate that the tic

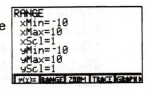

marks on the axes are one unit apart. The values entered on this screen may be changed by using the cursor arrows to move to the desired line and typing over the existing entry. When drawing a graph, you may set the desired viewing window on the calculator as well as scale the X-axis and Y-axis.

> **REMEMBER:** Every time the core unit indicates "press **[WINDOW]**" you will need to press **[GRAPH] [F2](RANGE)/ (WIND)**. When the core unit indicates "press **[GRAPH]**", you should press **[F5](GRAPH)** from the **GRAPH** menu.

Before proceeding further, press [GRAPH] [F1](y(x) =) and clear all entries.

☞ RETURN TO THE CORE UNIT AND COMPLETE THE EXERCISES ON PAGE 109.

ALTERING THE VIEWING WINDOW

The last unit addressed the size of the calculator screen (viewing window). Because the screen is 127 pixel points wide by 63 pixel points high, there are 126 horizontal spaces and 62 vertical spaces to light up. When tracing on a graph, the readout changes according to the size of the space. The size of the space can be controlled by the following formulas:

$$\frac{Xmax - Xmin}{126} = \text{horizontal space width}, \quad \frac{Ymax - Ymin}{62} = \text{vertical space height.}$$ We

will now examine some preset viewing windows and how they affect the pixel space size.

Press **[GRAPH]** **[F3](ZOOM)**. Pressing **[MORE]** repeatedly will display the remaining menu selections and then return the display to the original screen.

ZBOX	Boxes in and enlarges a designated area.
ZIN	Acts like a telephoto lens and "zooms in".
ZOUT	Acts like a wide-angle lens and "zooms out".
ZSTD	Automatically sets standard viewing window to [-10,10] by [-10,10].
ZPREV	Resets **RANGE** values to values used prior to the previous ZOOM operation.
ZFIT	Resets yMin and yMax on the **RANGE** screen to include the minimum and maximum y-values that occur between the current xMin and xMax settings.
ZSQR	"Squares up" the previously used viewing window.
ZTRIG	Sets the **RANGE** to built-in trig values.
ZDECM	Sets cursor moves to ONE tenth of a unit per move.
ZDATA	*TI-86 only: Automatically sets the viewing window to accomodate statistical data.*
ZRCL	Sets **RANGE** values to those stored by the user (see ZSTO).
ZFACT	Sets the zoom factors used in **ZIN** and **ZOUT**.
ZOOMX	Graph display is based on xFact only when zooming in or out.
ZOOMY	Graph display is based on yFact only when zooming in or out.
ZINT	Cursor moves are ONE integer unit per move.
ZSTO	Stores current **RANGE** values for future use. Values are recalled by **ZRCL**.

☞ Return to core unit pg.112 and begin reading at **Zdecimal**.

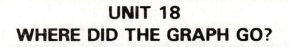

UNIT 18
WHERE DID THE GRAPH GO?

*Unit 17 is a prerequisite for this unit. Answers appear at the end of the unit.

Many times students are frustrated when the equation they have carefully keystroked into the **Y**= screen does not appear when **GRAPH** is pressed. What actually happens to the graph? Suppose you graphed $Y = 2X^2 + 4X + 12$ on graph paper and then graphed this same equation on the calculator with the viewing window set to ZStandard. The figure at the right illustrates the handsketched graph with the section displayed on the ZStandard screen outlined in a bold black line. The viewing window selected is not large enough to display the graph. This unit will give you the practice necessary to feel confident about setting the viewing window correctly to display a complete graph. **A complete graph is defined to be one in which all x- and y-intercepts are visible as well as any peaks/maximums and valleys/minimums.**

Before proceeding, press **[ZOOM] [6:ZStandard]** (TI-85/86 USERS PRESS **[GRAPH] [F3](ZOOM) [F4](ZSTD)**) to set the standard viewing WINDOW.) Enter 4X - 18 on the **Y**= screen and press **[GRAPH]**. The graph at the right should be displayed. The X-intercept is visible, but the Y-intercept is not.

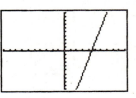

If the equation entered is not displayed on the graphing screen, the first item to be checked is the entry of the equation on the **Y**= screen. Is the equation *SELECTED* to be graphed? That is, is the equal symbol highlighted? If it is, proceed. If it is not, move the cursor over the equal sign and press **[ENTER]** to highlight the equal sign, thus activating the equation.

TI-85/86	TI-85/86 USERS SHOULD REMEMBER THAT **[F5](SELCT)** ON THE **Y(x)**= MENU WILL BE USED TO ACTIVATE AND DEACTIVATE EQUATIONS.

If the equation is activated, then begin the process of adjusting the WINDOW by locating the X and Y-intercepts of the graph.

Press **< 2nd >[Calc] [1:value]** to access the value option. Enter 0 at the X= prompt (because y-intercepts have an x-value of 0) and press **[ENTER]**. At the bottom of the screen the cursor's location (the y-intercept) of X = 0 and Y = -18 is displayed.

TI-85/86	WITH THE **GRAPH** MENU DISPLAYED, PRESS **[MORE] [MORE]** AND THE APPROPRIATE F KEY FOR **EVAL**.

Exit the GRAPH screen and enter the WINDOW screen by pressing **[WINDOW]**.

TI-85/86	PRESS **[F2](RANGE)**, WHICH IS (WIND) ON THE TI-86.

Use the down arrow key to move the cursor to Ymin and replace the Ymin with -20, a value smaller than -18. This smaller value allows a clear view of the Y-intercept.
Pressing **[GRAPH]** displays the screen at the right. This a satisfactory graph because both the X and Y-intercepts are displayed.

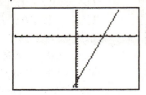

Reset the WINDOW to the standard viewing window.

Example 1: Graph $y = x^3 - 15x^2 + 26x$ in an appropriate viewing window.

Solution: Press **[Y =]**, enter $X^3 - 15X^2 + 26X$ at Y1 = and press **[GRAPH]**. A satisfactory graph of this equation should display all of its interesting features. Press **[TRACE]** (TI-85/86 users press **[F4] (TRACE)**) and TRACE along the graph to the right (using the right arrow key) and record the X and Y-intercepts as encountered. Remember, you are not on the ZInteger or ZDecimal screens. The X-intercepts may only be close approximations with the TRACE feature. (HINT: This is a third degree equation. There could be three X-intercepts.) Each of the screens below indicate the points closest to the X-intercepts that you should be able to locate.

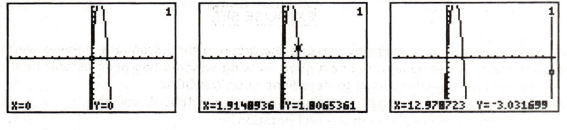

| X and Y-intercept | X-intercept ≈ 2 | X-intercept ≈ 13 |

Are all the peaks (maximums) and valleys (minimums) of the graph displayed? A satisfactory viewing window is a window that includes the X and Y-intercepts as well as all the peaks and valleys of the graph. TRACE the curve again, going left this time, to determine the lowest **Y value** (valley/minimum) and the highest **Y value** (peak/maximum) displayed (to the nearest whole number value).

valley/minimum = _____   peak/maximum = _____

Press **[WINDOW]** and adjust the Max and Min values for both X and Y to include the intercept points on the X-axis and the maximum (y = 12) and minimum (y = -252) Y values. It is suggested that you enter values that are a few units larger (or smaller) than the intercepts, maximum, and minimum you recorded. Suggested values are:
Xmin = -2   Ymin = -255
Xmax = 15   Ymax = 15

The graph is displayed using the above WINDOW values. The tic marks have been adjusted by setting the Xscl = 1 and the Yscl = 20. You may prefer to set both scales equal to zero and eliminate all tic marks.

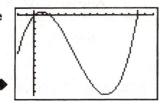

PROCEDURE FOR OBTAINING GOOD VIEWING WINDOWS

1. BEGIN BY DISPLAYING THE GRAPH IN THE STANDARD VIEWING WINDOW.
2. USE THE VALUE (EVAL) FEATURE TO DISPLAY THE COORDINATES OF THE Y-INTERCEPT. IF THE INTERCEPT IS OUT OF THE STANDARD VIEWING WINDOW, ADJUST THE WINDOW TO ACCOMMODATE THE INTERCEPT AND RETURN TO THE GRAPH SCREEN.
3. WITH THE TRACE FEATURE ACTIVATED, TRACE FIRST RIGHT AND THEN LEFT, RECORDING THE FOLLOWING INFORMATION AS ENCOUNTERED:
   A. X-INTERCEPT(S) - RECALL, THIS IS WHERE THE Y-COORDINATE IS ZERO. IF YOU CANNOT TRACE TO THIS EXACT POINT, REMEMBER, WHEN THE Y-VALUE CHANGES SIGN, YOU ARE CLOSE TO THE ZERO AND CAN SET THE X-VALUE APPROPRIATELY ON THE WINDOW SCREEN.
   B. AS YOU ARE TRACING TO FIND X-INTERCEPTS, RECORD ANY MAXIMUM OR MINIMUM Y-VALUES ENCOUNTERED OR USE THE VALUE (EVAL) FEATURE. SET THE WINDOW VALUES TO INCLUDE THESE "INTERESTING" FEATURES. REMEMBER TO ALWAYS SET THE WINDOW VALUES SLIGHTLY LARGER THAN THE MAXIMUM OR SMALLER THAT THE MINIMUM VALUES YOU ENCOUNTERED.

As your algebraic knowledge increases, you will be able to blend this knowledge with calculator expertise to expedite the process of determining a good viewing window.

EXERCISE SET

Directions: Begin each problem by viewing the graph in the ZStandard viewing window. Sketch a graph of the equation once a good viewing window has been established.
   a. Record the values used to determine your WINDOW.
   b. Sketch the graph that is displayed for these WINDOW values.
   c. Record the WINDOW in interval notation.

1. $Y = -2X^2 + 19$

    Y-intercept = _____   Ymax = _____

    X-intercepts = _____, and _____

    [_____,_____] by [_____,_____]
     Xmin   Xmax   Ymin   Ymax

2. $Y = (1/2)X^4 - 10X^2 + 25$

    Y-intercept = _____

    X-intercepts = _____, _____, _____, _____

    Ymax = _____   Ymin = _____, _____

    [_____,_____] by [_____,_____]
     Xmin   Xmax   Ymin   Ymax

121

3.　　　$Y = \sqrt[3]{X^2 - 5X - 300}$

(If you do not remember how to enter radicals other than square roots, refer to Unit 5.)

Y-intercept = _____

X-intercepts = _____, _____,

Ymin = _____

$$[\underline{\hspace{1.5cm}}, \underline{\hspace{1.5cm}}] \text{ by } [\underline{\hspace{1.5cm}}, \underline{\hspace{1.5cm}}]$$
　　Xmin　　Xmax　　　Ymin　　Ymax

4.　　　a. Establish an appropriate viewing window for the equation $Y = 0.1X^4 - 8X^2 + 5X + 1$ and sketch the result.

$$[\underline{\hspace{2cm}}, \underline{\hspace{1.5cm}}] \text{ by } [\underline{\hspace{1.5cm}}, \underline{\hspace{1.5cm}}]$$

b. Using the space width formulas discussed in the unit entitled "Preparing to Graph", determine the missing window values if each move of the TRACE cursor is to be **ONE** unit.

[-45, X] by [-200,10]　　　　　X = _____

Sketch the display provided by this window.

✍ c. Space width formulas can be used for both coordinates and yet when tracing, the Y-coordinates displayed are not consistently integers. Explain.

Solutions: 1. Y-intercept:19, Ymax: at least 20, X-intercepts: approximately -3 and 3; [-4, 4] by [-10, 20]

2. Y-intercept:25, X-intercepts: approximately -4, -1.7, 1.7, 4, Ymax:25, Ymin:-26, -26; [-5, 5] by [-30, 30]

3. Y-intercept: approximately -7, X-intercepts:-15,20, Ymin: -7; [-25, 30] by [-10, 10].

4. a. [-10, 10] by [-200, 10]　　　　b. [-45, 49] by [-200,10]　　　　c. Y-values are determined by the x-values that are selected to replace the variables in the equation.

122

UNIT 19
FUNCTIONS

*Unit17 is a prerequisite for this unit. Answers appear at the end of this unit.

A function is a collection of ordered pairs in which each *x*-value is paired with a unique *y*-value.

Domain and Range

The set of all *acceptable* x-values of the ordered pairs in a function is called the **domain**. The set of all resultant y-values is called the **range**. When determining domains and ranges of functions, bear in mind that there are three tools: the algebraic definition of the function, the table feature of the graphing calculator (82/83/83plus/86 users) and the graph.

TI-85	TI-85 USERS SHOULD BE CAREFUL NOT TO CONFUSE THE RANGE OF A FUNCTION (THE SET OF ALL Y VALUES) WITH THE **RANGE** SCREEN WHERE PARAMETERS ARE DETERMINED FOR THE VIEWING WINDOW.

EXERCISE SET

Directions: Use the **TRACE** feature to determine the domain and range of each of the following functions. TRACE along the path of the graph and examine the X and Y values displayed at the bottom of the screen. These will be of assistance in determining the domain and range. You may use any viewing WINDOW you desire, however you may discover that some viewing WINDOWS are more "informative" than others. Suggestion: view and TRACE on each graph in each of the following WINDOWS - ZStandard, ZDecimal, ZInteger before determining the domain. (TI-85/86 users recall that these preset windows are listed under the **ZOOM** menu as ZSTD, ZDECM, and ZINT.)

1. $Y = X^2 + 1$ Domain = _____

   Range = _____

2. $Y = 3X + 5$ Domain = _____

   Range = _____

3. $Y = \sqrt{X + 5}$ Domain = _____

   Range = _____

4. $Y = X^3 + 4X^2 + 2$ Domain = _____

   Range = _____

✍ 5. Consider the two functions Y1 = X and Y2 = $\sqrt{X} \cdot \sqrt{X}$ and respond to each of the following questions.

a. Would you expect the two domains to be the same? Why or why not?

b. Considering only the algebraic equations (no calculator allowed), what should the domains should be?

c. How can the TABLE feature help in determining domain?

d. What does the graph of each function indicate the domain should be? Is one viewing window more informative than another?

✍ 6. Consider the function $Y = \dfrac{(2X + 3)(X + 2)}{2X + 3}$ and respond to each of the following questions.

a. What do you expect the domain to be?

b. Considering only the algebraic equation (no calculator allowed), what should the domain should be?

c. How can the TABLE feature help in determining domain?

d. What does the graph of the function indicate the domain should be? Is one viewing window more informative than another?

When an equation is graphed, the VERTICAL LINE TEST can be used to determine if the equation is a function. Remember, to be a function there must be only one Y- value for each X-value. Thus, when a vertical line is passed across the graph (from left to right) the line will intersect the graph in only one point at a time if the graph represents a function.

Graph the equation $Y = X^2 + 2X + 2$ and display the graph on the standard viewing WINDOW. To "draw" a vertical line with the calculator, press **[2nd]** **<DRAW>**, **[4:Vertical]** (i.e. vertical line).

TI-85/86	TO ACCESS THE VERTICAL LINE, PRESS **[GRAPH]** **[MORE]** **[F2](DRAW)** **[F3](VERT)**.

The vertical line is actually displayed on the Y-axis initially. Press the **[►]** and **[◄]** arrow keys to move the vertical line right and/or left across the graph. Since the vertical line does not intersect the graph in more than one place at a time the equation represents a function. Using the **[►]** or **[◄]** keys, return the vertical line to the Y-axis so that it is no longer visible. Each time you graph a new equation you must go back to the **DRAW** menu to access the Vertical Line option (as well as other options in this menu). Note: Anything that is DRAWN on the graph via the **DRAW** menu can only be cleared by pressing **[1:ClrDraw]**.

TI-85/86	ACCESS THE **DRAW** MENU, PRESS **[MORE]** UNTIL **(CLDRW)** FOR CLEAR DRAW IS DISPLAYED, THEN PRESS THE APPROPRIATE **F** KEY.

EXERCISE SET CONTINUED

Directions: Use the Vertical Line option from the **DRAW** menu on the graph of each of the following equations to decide if the graph of the equation represents a function. Sketch the graph displayed **AND** sketch the vertical line at some point on the graph.

7. $Y = -X^2 - 8X - 10$

   Function? (yes or no)_____

8. $Y = \frac{1}{2}X^3 + X^2 - 2X + 1$

   Function? (yes or no)_____

9. $Y = 2\sqrt{X + 4}$

   Function? (yes or no)_____

Horizontal Line Test

The horizontal line test is used on the graph of a function to determine if it is one-to-one. A function is one-to-one if for each Y-value in the range there is one and only one X-value in the domain. A horizontal line passed across the graph will not intersect the graph in more than one place at a time if the graph is that of a one-to-one function.

TI-85/86	THE TI-85 DOES NOT HAVE A HORIZONTAL LINE OPTION. USE A STRAIGHT EDGE TO PERFORM THE HORIZONTAL LINE TEST ON THE EXAMPLE AND THE EXERCISES. THE TI-86 HAS THE HORIZONTAL LINE OPTION. IT IS LOCATED NEXT TO THE VERT ON THE DRAW MENU.

Re-enter $Y = X^2 + 2X + 2$ and display the graph. Press **[2nd]** **<DRAW>** **[3:Horizontal]**. Use the **[▲]** and **[▼]** arrow keys to move the horizontal line up and down the graph. The horizontal line is originally positioned on the X-axis. When you are finished with this option, return the horizontal line to its beginning position on the X-axis. This is not the graph of a one-to-one function because the horizontal line intersects the graph in <u>two</u> <u>places</u> everywhere except at the vertex.

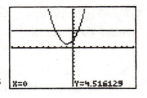

EXERCISE SET CONTINUED

Directions: Use the Horizontal Line option from the **DRAW** menu on the graph of each of the functions to decide if the graph represents a one-to-one function. Sketch the graph displayed and the horizontal line at a location where it intersects the function in more than one place. If the function is one-to-one do not draw in the horizontal line.

10. $Y = -X^2 - 8X - 10$

   1-1 Function? (yes or no)_____

11. $Y = \frac{1}{2}X^3 + X^2 - 2X + 1$

   1-1 Function? (yes or no)_____

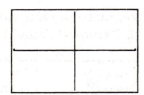

12. $Y = 2\sqrt{X + 4}$

   1-1 Function? (yes or no)_____

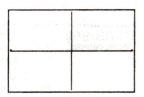

INVERSES

If a function is one-to-one, i.e. passes both the vertical and horizontal line tests, it will have an inverse function.

Consider the function Y = 2X + 3. Enter 2X + 3 at the Y1 = prompt on the calculator. Press **[ZOOM] [4:ZDecimal] [WINDOW]** and double all window values except for Xscl and Yscl. This will be referred to in the future as the **ZDecimal x 2** viewing window. Copy the display. This is a one-to-one function because the graph passes both the vertical and horizontal line tests.

TI-85/86	PRESS **[F3](ZOOM) [MORE] [F4](ZDECM)** AND DOUBLE ALL RANGE VALUES EXCEPT XScl AND yScl. THIS ZDECM x 2 SCREEN IS APPROXIMATELY THREE UNITS LONGER AT EACH END OF THE X AXIS THAN THE SCREEN THAT IS DISPLAYED FOR THE TI-82/83/83PLUS. HOWEVER, THIS SHOULD NOT AFFECT THE RECORDING OF THE CALCULATOR DISPLAY TO THE SCREENS GIVEN.

Algebraically find the inverse below by (a) interchanging the X and Y variables and (b) solving for Y.

The inverse equation should be $Y = \frac{1}{2}X - \frac{3}{2}$. Enter this equation at the Y2 = prompt. At the Y3 = prompt, enter X to graph Y = X. Your display of all three functions should look like the one at the right.

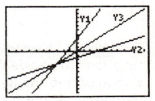

Y1 and Y2 are symmetric across the line Y = X (the Y3 line). This will always be true of a function and its inverse.

To use the calculator to **DRAW** the inverse function, the function must be entered on the **Y =** screen. **Since the function is entered at Y1, go back and delete Y2 and Y3 before proceeding.** Press **[2nd] <QUIT>** to return to the home screen.

Instruct the calculator to **DRAW** the inverse of Y1 (DrawInv): Press **[2nd] <DRAW> [8:DrawInv] [Vars] [▶]**(to highlight Y-Vars) **[1:Function...] [1:Y1] [ENTER]**.

(TI-82)	Press **[2nd] <DRAW> [8:DrawInv]**. To enter **Y1** after the draw inverse command, press **[2nd] <Y-vars> [1:Function] [1:Y1] [ENTER]**.

TI-85/86	THE DRAW INVERSE COMMAND IS ACCESSED FROM EITHER THE HOME SCREEN OR THE GRAPH SCREEN BY PRESSING **[GRAPH] [MORE] [F2](DRAW)** AND PRESSING **[MORE]** UNTIL (DRINV) IS DISPLAYED. PRESS THE APPROPRIATE F KEY TO ACCESS THE COMMAND. DISPLAYED ON THE HOME SCREEN IS THE DRINV COMMAND. ENTER y1 AFTER THE COMMAND BY PRESSING **[2ND] [ALPHA] <y> [1] [ENTER]**. REMEMBER: THE **y** MUST BE LOWER CASE!

Y1 will be graphed first and the INVERSE of Y1 will be drawn second. The inverse that is drawn is the same line as the one graphed at Y2, however, because it is drawn (and not graphed from the Y = screen) you will not be able to interact with the graph. That is to say, you will not be able to TRACE, use INTERSECT, ROOT/ZERO, VALUE, TABLE, etc.

Standard function notation is **f(X)**, where f denotes the function, X is the independent variable and f(X) represents the function's value at X. If an equation represents the graph of a function, then the Y variable in the equation may be replaced by the f(X) notation.

In Exercise 7, $Y = -X^2 - 8X - 10$ was determined to be a function. The equation can now be written as $f(X) = -X^2 - 8X - 10$. Since the calculator will only graph functions, Y1 (or y1(x) on the TI-85/86) is equivalent to the function denoted as **f**.

To evaluate the function $f(X) = -X^2 - 8X - 10$ at $X = 2$, write: $f(2) = -(2)^2 - 8(2) - 10 = -30$. Thus when $X = 2$, $f(X) = -30$ (i.e. $Y = -30$). This would yield the ordered pair (2,-30) on the graph of f(X). Verify with the TABLE. Be sure it is incremented by 1 and enter $-X^2 - 8X - 10$ at the Y1= prompt. Find $X = 2$ in the table. When $X = 2$, the table indicates that Y1 = -30. You could also use the VALUE (EVAL) feature as discussed in Unit 10, "Applications of Quadratic Equations".

> **TI-85** SINCE THERE IS **NO TABLE** FEATURE, YOU SHOULD CONSIDER USING THE FUNCTION EVALUATION FEATURE. PRESS **[GRAPH]** **[F1]**(y(x)=) AND ENTER $-X^2 - 8X - 10$ AT THE y1(x)= PROMPT. PRESS **[GRAPH]** **[MORE]** **[MORE]** **[F1]EVAL**) AND ENTER THE DESIRED X VALUE AT THE PROMPT. THE **EVAL** OPTION IS RESTRICTED TO X VALUES BETWEEN xMIN AND xMAX ON THE **RANGE** SCREEN. ADJUST THE **RANGE** AS NECESSARY. IF MORE THAN ONE EQUATION IS ENTERED ON THE y(x)= SCREEN, EACH EQUATION WILL BE EVALUATED AND THE RESULTS FOR EACH CAN BE DISPLAYED BY PRESSING THE UP OR DOWN CURSOR ARROWS. THE EQUATION NUMBER IS DISPLAYED IN THE TOP RIGHT HAND CORNER OF THE SCREEN.

The TI-82/83/83plus/86 uses the equivalent function notation of Y(X) instead of f(X). To evaluate $f(X) = -X^2 - 8X - 10$ at $X = 2$, the f(X) expression must be entered on the **Y =** screen. Enter the polynomial at the Y1= prompt. Return to the home screen by pressing **[2nd]** <**QUIT**>. To compute the value for f(X) when $X = 2$, press **[VARS]** **[▶]** **[1:function]** **[1:Y1]** **[(]** **[2]** **[)]** and then **[ENTER]** to compute. (TI-86 users press **[2nd]** <**alpha**> <**y**> **[1]** **[(]** **[2]** **[)]**).

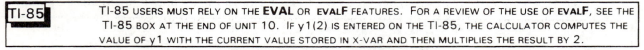

```
Y1(2)
            -30
```

> **TI-82** Find Y1 by pressing **[2nd]** <**VARS**> **[1:Function]** **[1:Y1]**.

The value of the function will be displayed for the X-value of 2. In ordered pair form this would be (2,-30).

> **TI-85** TI-85 USERS MUST RELY ON THE **EVAL** OR **evalF** FEATURES. FOR A REVIEW OF THE USE OF **evalF**, SEE THE TI-85 BOX AT THE END OF UNIT 10. IF y1(2) IS ENTERED ON THE TI-85, THE CALCULATOR COMPUTES THE VALUE OF y1 WITH THE CURRENT VALUE STORED IN X-VAR AND THEN MULTIPLIES THE RESULT BY 2.

EXERCISE SET CONTINUED

Directions: Enter each of the following functions on the **Y =** screen. Evaluate for the indicated value of X using the Y-VARS capability of the calculator (i.e. Y(X) notation). Copy the screen that displays the result. Record your final information as an ordered pair.

13.　Evaluate $Y = 3X^3 - 2X^2 + X - 5$ for $X = -\dfrac{3}{20}$.
　　Screen display:

　　Ordered pair:_____ (in decimal form)

　　Ordered pair:_____ (in fraction form)

14. Evaluate $Y = \sqrt{2X - 5}$ for $X = 3.5$.

Screen display:

Ordered pair:_____

15. Evaluate $Y = \dfrac{2X + 3}{X^2 + 4X - 5}$ for $X = -\dfrac{1}{2}$

Screen display:

Ordered pair:_____ (in fraction form)

16. Evaluate $Y = \sqrt{3X + 5}$ for $X = -8$.

Screen display:

Why did you get this display? Explain carefully.

17. The profit or loss for a publishing company on a textbook supplement can be represented by the function $f(X) = 10X - 15000$ (X is the number of supplements sold and $f(X)$ is the resulting profit or loss).
a. Use the Y-VARS capability to determine the amount of profit (or loss) if 2000 supplements are sold. Copy the screen display to justify your work.

b. How can you tell if the $5000 is profit or loss?

c. What would be the profit (or loss) if 1000 supplements are sold?
Copy the screen display to justify your work.

✐18. In your own words, explain the various approaches for determining domain and range of functions. Include both algebraic and calculator methods in your discussion.

✐19. In your own words, explain the similarities and differences between a function and a one to one function. Your discussion should not be limited to the vertical and horizontal line tests.

Solutions: 1. domain = $\mathbb{R}$, range = $\{Y \mid Y \geq 1\}$ **2.** domain = $\mathbb{R}$, range = $\mathbb{R}$

3. domain = $\{X \mid X \geq -5\}$, range = $\{Y \mid Y \geq 0\}$ **4.** domain = $\mathbb{R}$, range = $\mathbb{R}$

5. a. answers may vary   **b.** $Y = x$ should have a domain of all real numbers and $y = \sqrt{x}\sqrt{x}$ should have a domain of all non-negative real numbers.   **c.** The table agrees with the statement made in part b.   **d.** The graph agrees with the statement made in part b. With respect to viewing windows, answers may vary.

6. a. The domain should be $\{x \mid x \neq -3/2\}$.   **b.** The domain is $\{x \mid x \neq -3/2\}$.   **c.** The table will not support the declared domain unless you know to increment the table by ½. The graph supports the domain stated in part b. With respect to viewing windows, answers may vary.

7. Yes   **8.** Yes   **9.** Yes   **10.** No   **11.** No   **12.** Yes

13. (.15, -5.205125), (-3/20, -41641/8000)   **14.** (3.5, 1.414213562)

15. (-1/2, -8/27)   **16.** $\sqrt{3X + 5}$ is equivalent to $\sqrt{-19}$ when $X = -8$. The square root function is undefined for negative radicands.

17. a. $Y_1(2000)$ 5000   **b.** The $5,000 is positive, and therefore a profit.
c. $Y_1(1000)$ -5000   The negative indicates that the $5,000 would be a loss.
18. answers may vary

UNIT 20
DISCOVERING PARABOLAS

*Unit 17 is a prerequisite for this unit. Answers appear at the end of the unit.

This unit explores the graphs of quadratic equations, specifically quadratic functions. A quadratic function is an equation in the form $y = ax^2 + bx + c$, where a, b, and c are real numbers. These values (a,b,c) will affect the size and location of the curve. This unit is an exercise in discovery. As each equation is graphed, study the size and location of the parabola and compare this information to the coefficients in the equation.

1.   Set the viewing window to ZDecimal x 2, [-9.4,9.4] by [-6.2,6.2], with both scales equal to 1. (The TI-85/86 RANGE/WIND values will be [-12.6,12.6] by [-6.2,6.2].) The TRACE feature will be used to help discover some of the characteristics of these parabolas. This viewing window was selected because the cursor moves will be in tenths of units.

2.   The vertex is the minimum point (or maximum point) on the graph of a parabolic curve. GRAPH and TRACE to find the vertex of each of the parabolas graphed by the given equation. Sketch the graph and record the coordinates of the vertex as ordered pairs.

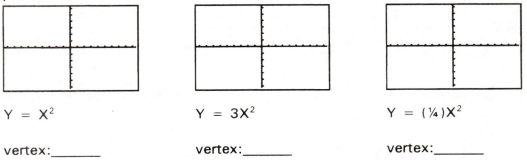

Y = X^2 Y = 3X^2 Y = (¼)X^2

vertex:_____   vertex:_____   vertex:_____

✎3.   Observation: What effect does the coefficient on the X^2 term have on the graph of the equation?

4.   GRAPH and TRACE to find the vertex of the following parabolas. Sketch the graph and record the coordinates of the vertex.

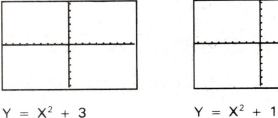
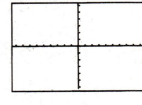
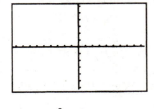

Y = X^2 + 3   Y = X^2 + 1   Y = X^2 - 2

vertex:_____   vertex:_____   vertex:_____

✎5.   Observation: What effect does the constant term have on the graph of the equation?

6.	GRAPH and TRACE to find the vertex of the following parabolas. Sketch the graph and record the coordinates of the vertex.

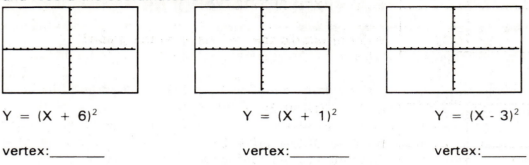

$Y = (X + 6)^2$

vertex:_____

$Y = (X + 1)^2$

vertex:_____

$Y = (X - 3)^2$

vertex:_____

✎7.	Observation: When a value is added or subtracted to the X <u>before</u> the quantity is squared, what effect does it have on the graph?

8.	GRAPH and TRACE to find the vertex of the following parabolas. Sketch the graph and record the coordinates of the vertex.

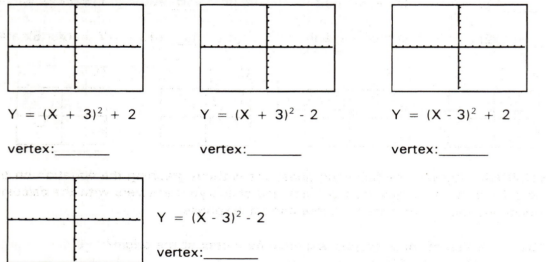

$Y = (X + 3)^2 + 2$

vertex:_____

$Y = (X + 3)^2 - 2$

vertex:_____

$Y = (X - 3)^2 + 2$

vertex:_____

$Y = (X - 3)^2 - 2$

vertex:_____

9.	Based on your observations from the previous problems, what should the vertex of the parabola $Y = (X + 115)^2 - 38$ be? **DO NOT ATTEMPT TO ANSWER THIS QUESTION BY GRAPHING THE EQUATION!** Use the information gathered in the previous problems to determine the vertex.

_____

10.	Compare each pair of equations by graphing on the same graph screen, ZDecimal x 2.

a. $Y = 3(X + 2)^2 + 2$

b. $Y = (X + 2)^2 + 2$

What effect did the "3" have on the graph?

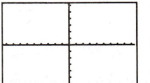

a. $Y = (¼)(X + 2)^2 + 2$

b. $Y = (X + 2)^2 + 2$

What effect did the "¼" have on the graph?

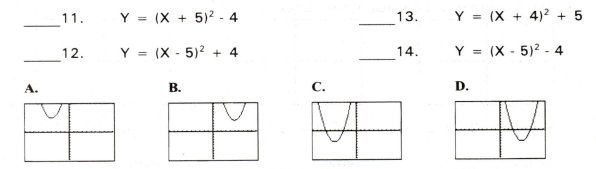

a. $Y = 3(X + 2)^2 + 2$

b. $Y = -3(X + 2)^2 + 2$

What effect did the negative sign on the 3 have on the graph?

Using what you have learned in this unit, match each graph below with one of the equations #11-14. DO NOT ENTER ANY EXPRESSIONS ON THE CALCULATOR.

_____11. $Y = (X + 5)^2 - 4$ _____13. $Y = (X + 4)^2 + 5$

_____12. $Y = (X - 5)^2 + 4$ _____14. $Y = (X - 5)^2 - 4$

A.   **B.**   **C.**   **D.**

CONCLUSIONS: Answer the following questions without graphing the equation on the graphing calculator. You may then go back and check your answers with the calculator. Some questions may have more than one correct response.

_____15.   Which of these graphs will have its vertex at the origin?
a. $Y = (X - 5)^2$
b. $Y = X^2$
c. $Y = (4/5)X^2$
d. $Y = 2X^2 + 7$
e. $Y = 4(X + 3)^2 - 7$

_____16.   Which of these is the graph of $Y = X^2$ translated (shifted) two units to the left of the Y-axis?
a. $Y = 2X^2$
b. $Y = (X + 2)^2$
c. $Y = X^2 - 2$
d. $Y = (X + 2)^2 - 2$
e. $Y = (X - 2)^2$

_____17. Which of these is the graph of $Y = X^2$ translated 2 units down from the X-axis?

a. $Y = 2X^2$
b. $Y = (X + 2)^2$
c. $Y = X^2 - 2$
d. $Y = (X + 2)^2 - 2$
e. $Y = (X - 2)^2$

_____18. Which of these graphs has a maximum point?

a. $Y = 2X^2$
b. $Y = -2(X - 2)^2$
c. $Y = X^2 - 2$
d. $Y = (X + 2)^2 - 2$
e. $Y = 2 - X^2$

✎19. a. Based on your observations, for the parabola $y = a(x - h)^2 + k$, what effect does the sign of "a" have on the orientation of the parabola?

b. As $|a|$ increases, what affect does it have on the size of the parabola?

c. The value of h will shift (translate) the parabola which direction?

d. The value of k will shift (translate) the parabola which direction?

20. Consider the quadratic equation $Y = 2X^2 - 12X + 19$. Enter this equation on the calculator and sketch the graph.

a. Access the **CALC** menu and select [**3:minimum**] (or 4:maximum as appropriate) to find the vertex of the parabola. The vertex appears to be (3,1). Confirm this by setting the TABLE start/min to 0, and accessing the TABLE. Notice that coordinates are symmetric on either side of (3,1).

TI-85/86	FMIN AND FMAX ARE LOCATED UNDER THE MATH SUBMENU OF THE GRAPH MENU OF BOTH OF THESE CALCULATORS. TI-85 USERS ONLY: SINCE THERE IS **NO TABLE** FEATURE, YOU SHOULD CONSIDER USING THE **EVAL** OR **EVALF** FEATURES. FOR A REVIEW OF THE USE OF **EVAL OR EVALF**, SEE THE TI-85 NOTE AT THE END OF UNIT 10.

NOTE: When using the graph screen to determine coordinates, you should be aware that the display coordinate values approximate the actual mathematical coordinates. The accuracy of these display values is determined by the height and width of the pixel space being displayed. The space height/width formulas are discussed in detail in Unit 17 entitled, "Preparing to Graph: Calculator Viewing Windows.

    b. Previously the vertex of the parabola could be read from the equation that was in the form $y = a(x - h)^2 + k$. However, when the equation is in the form $y = ax^2 + bx + c$ you must complete the square on x to be able to read the vertex.

    c. Since the vertex is at (3,1), then we can expect the equation to have the form $y = a(x - 3)^2 + 1$. The only value we do not know is "a". Can you predict the value of **a**?_____

    d. To complete the square on X of $Y = 2X^2 - 12X + 19$, the first step would be to factor out the coefficient of X^2, i.e. $Y = 2(X^2 - 6X + \underline{\quad}) + 19$.

Continuing this process produces:
$Y = 2(X^2 - 6X + \underline{9}) + 19 - \underline{2(9)}$
$Y = 2(X - 3)^2 + 1$

Notice **a** = 2. Was your prediction correct? Will **a** always be equal to the coefficient of the x^2 term? Why or why not?

21.   a. Graph the equation $Y = 4X^2 + 8X - 2$. Sketch the display.

    b. Use the "minimum" option of the CALC menu to find the vertex of the parabola. Record the vertex coordinates:

    (_____ , _____)

    c. Using what you learned in #20, write the equation of the parabola $Y = 4X^2 + 8X - 2$ in $y = a(x - h)^2 + k$ form:_____

    d. Complete the square on $Y = 4X^2 + 8X - 2$ to write the equation in $y = a(x - h)^2 + k$ form to be sure that the result agrees with your response in letter **c**.

Solutions: 2. (0,0), (0,0), (0,0) **3.** It affects the width of the parabola. **4.** (0,3), (0,1), (0,-2) **5.** It

moves the parabola up or down the Y-axis. **6.** (-6,0), (-1,0), (3,0) **7.** It shifts the parabola left or right

along the X-axis. **8.** (-3,2), (-3,-2), (3,2), (3,-2) **9.** (-115,-38) **10.** The "3" made the graph

"steeper". The "¼" made the graph "flatter". The negative sign turned the graph "upside down".

11. c **12.** b **13.** a **14.** d **15.** b,c **16.** b,d **17.** c,d **18.** b,e **19a.** If "a" is positive, the parabola

opens up and has a minimum. If "a" is negative , the parabola opens down and has a maximum.

19b. |a| determines the width of the parabola. **19c.** "h" shifts the parabola right or left.

19d. "k" shifts the parabola up or down. **21b.** (-1,-6)   **21c.** $y = 4(x + 1)^2 - 6$

UNIT 21
TRANSLATING AND STRETCHING GRAPHS

*Unit 17 is a prerequisite for this unit. Answers appear at the end of the unit.

| Translations |

This section examines how the graphs of elementary functions can be shifted vertically and horizontally in the coordinate plane. These shifts, called *translations*, may be either horizontal, vertical or both. The unit "Discovering Parabolas" was an in-depth look at the graphs of parabolic functions and their corresponding equations. You may want to complete the unit "Discovering Parabolas" before working on this unit; however, it is not a prerequisite.

Before proceeding, set the viewing WINDOW to a standard viewing window.

| EXERCISE SET |

1. a. Each of the following equations is in the form
    $y = f(x) + c$. The reference graph $f(x) = X^2$ is displayed.
    Graph $Y = X^2 + 4$ and $Y = X^2 - 4$ on this same set of
    axes, labeling each graph.
    The range of $Y = X^2 + 4$ is _____.
    The range of $Y = X^2 - 4$ is _____.
    The domain for both graphs is $\mathbb{R}$.

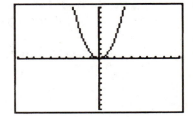

   b. The next group of graphs is in the form $y = f(x + b)$.
    The reference graph of $f(x) = X^2$ is displayed.
    Graph $Y = (X + 5)^2$ and $Y = (X - 5)^2$ on this same set of
    axes, labeling each graph.
    The domain for both graphs is _____ and the
    range for both graphs is _____.

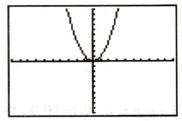

   ✍ c. CONCLUSION: In the general form
    $y = f(x + b) + c$, horizontal shifts result from changes in the variable _____ and
    vertical shifts result from changes in the variable _____.

   d. In the case of a parabola, does the horizontal shift ever affect the domain or
    range?

   e. Does the vertical shift affect the domain or range?

f. Based on the above conclusions, the graph of y = (x - 28)² + 12 should translate vertically _____ units _____ (up/down) and horizontally _____ units _____ (left/right) when compared to f(x) = x².
This would locate the vertex of the parabola at the coordinates (_____ , _____).

2. a. Using the same viewing WINDOW, consider the following absolute value equations in the form y = f(x) + c. The graph of f(x) = |X| has been graphed as the reference graph. Graph Y = |X| + 4 and Y = |X| - 4 on this same set of axes, labeling each graph.
The range of Y = |X| + 4 is _____.
The range of Y = |X| - 4 is _____.
The domain for both graphs is ℝ.

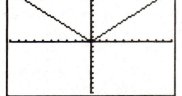

b. The next group of graphs is in the form y = f(x + b). The reference graph of f(x) = |X| is already displayed. Graph Y = |X + 5| and Y = |X - 5| on this same set of axes, labeling each graph.
The domain for both graphs is _____ and the range for both graphs is _____.

✍ c. CONCLUSION: In the general form y = f(x + b) + c, horizontal shifts result from changes in the variable _____ and vertical shifts result from changes in the variable _____ .

d. In the case of an absolute value function, does the horizontal shift ever affect the domain or range?

e. Does the vertical shift affect the domain or range?

f. Based on your conclusions, the graph of y = |x + 32| - 42 should translate vertically _____ units _____ (up/down) and horizontally _____ units _____ (left/right) when compared to the graph of f(x) = |x|.
This would locate the vertex of the absolute value function at the coordinates (_____ , _____).

3. a. Using the same viewing WINDOW, consider the following square root functions in the form y = f(x) + c. The graph of f(x) = √X has been graphed as the reference graph. Graph Y = √X + 4 and Y = √X - 4 on this same set of axes, labeling each graph.
The range of Y = √X + 4 is _____.
The range of Y = √X - 4 is _____.
The domain of both graphs is _____.

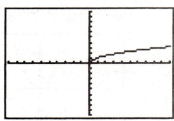

138

b. The next group of graphs is in the form y = f(x + b).
The reference graph of f(x) = $\sqrt{X}$ is displayed.
Graph Y = $\sqrt{X + 5}$ and Y = $\sqrt{X - 5}$ on this same set of
axes, labeling each graph.
The domain of Y = $\sqrt{X + 5}$ is _____ .
The domain of Y = $\sqrt{X - 5}$ is _____ .
The range for both graphs is _____ .

c. CONCLUSION: In the general form y = f(x + b) + c, horizontal shifts result from
changes in the variable _____ and vertical shifts result from changes in the variable
_____ .

d. In the case of a square root function, does the horizontal shift ever affect the
domain or range?

e. Does the vertical shift affect the domain or range?

f. Based on your conclusions, the graph of y = $\sqrt{x + 23}$ – 31 should translate
horizontally _____ units _____ (left/right) and vertically _____ units _____
(up/down) when compared to f(x) = $\sqrt{x}$.
This would locate the initial point of the square root curve at the coordinates
(_____ , _____).

Stretches, Compressions, and Reflections

Unlike translations, when a graph is stretched or compressed the domain and/or range are
not affected. The following exercises explore the effects of constant coefficients on the
graphs of functions. We will continue to use a standard viewing WINDOW.

4.   Graph each of the following groups of graphs that are in the form y = a · f(x). The
       reference graph, listed first in the series, is graphed for you. Label each graph on the
       display.

Y = X², Y = 4X², Y = (½)X²   Y = $|X|$, Y = 4$|X|$, Y = (½)$|X|$

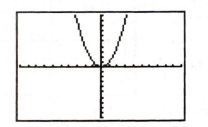   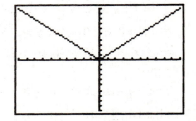

139

$Y = \sqrt{X}$, $Y = 4\sqrt{X}$, $Y = (\frac{1}{2})\sqrt{X}$

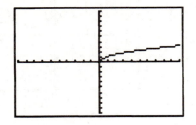

a.   In general, if a > 0, what effect does **a** have on the graph of $y = a \cdot f(x)$?

b.   What will happen to the graph of $y = a \cdot f(x)$ if a < 0? (If need be, go back to the previous problem and graph with values of *a* less than 0, (negative).

5.   The graph of $f(x) = \sqrt{x}$ swings right and up, $f(x) = -\sqrt{x}$ swings right and down. What must be done to f(x) so that the graph of the square root function swings left?

6.   Write the equation of a square root function whose domain is $\{X \mid X \leq 1\}$ and whose range is $\{Y \mid Y \geq -3\}$.

7.   Write the equation of a square root function whose domain is $\{X \mid X \leq -1\}$ and whose range is $\{Y \mid Y \leq 3\}$.

8.   Write the equation of a square root function whose domain is $\{X \mid X \geq 4\}$, whose range is $\{Y \mid Y \leq 5\}$.

9.   Write the equation of an absolute value function whose domain is $\mathbb{R}$, whose range is $\{Y \mid Y \leq 5\}$.

SOLUTIONS: 1. a. range of $Y = X^2 + 4$: $Y \geq 4$, range of $Y = X^2 - 4$: $Y \geq -4$ **b.** In the next group of graphs the domain is $\mathbb{R}$ and the range is $Y \geq 0$ for both graphs. **c.** Conclusion: "b" affects horizontal shift and "c" affects vertical shift. **d.** No **e.** The vertical shift affects only the range.
f. Thus $y = (x-28)^2 + 12$ translates vertically 12 units and horizontally 28 units with the vertex at (28,12).

2. a. range of $Y = |X| + 4$: $Y \geq 4$, range of $Y = |X| - 4$: $Y \geq -4$ **b.** In the next group of graphs the domain is $\mathbb{R}$ and the range is $Y \geq 0$ for both graphs. **c.** Conclusion: "b" affects horizontal shift and "c" affects vertical shift. **d.** No **e.** The vertical shift affects only the range.
f. Thus $y = |x+32| - 42$ translates vertically 42 units down and horizontally 32 units left with the vertex at (-32, - 42).

3. a. range of $Y = \sqrt{X} + 4$: $Y \geq 4$, range of $Y = \sqrt{X} - 4$: $Y \geq -4$. The domain of both graphs is $X \geq 0$.
b. In the next group of graphs the domain of $Y = \sqrt{X + 5}$ is $X \geq -5$ and the domain of $Y = \sqrt{X - 5}$ is $X \geq 5$. The range is $Y \geq 0$ for both graphs. **c.** Conclusion: "b" affects horizontal shift and "c" affects vertical shift. **d.** Only the domain is affected. **e.** Only the range is affected.
f. Thus $y = \sqrt{X + 23} - 31$ translates horizontally 23 units left and vertically 31 units down with the initial point of the curve at (- 23,- 31).

4. a. "a" affects the "width" of the graph. **b.** If $a < 0$ the graph will be symetric across the X-axis to the graph whose "a" is positive.

5. The opposite of the radicand must be graphed. **6.** $f(x) = \sqrt{1 - x} - 3$

7. $f(x) = -\sqrt{-1 - x} - 3$ **8.** $f(x) = -\sqrt{x - 4} + 5$ **9.** $f(x) = -|x| + 5$

UNIT 22
EXPONENTIAL AND LOGARITHMIC FUNCTIONS

*Unit 19 is a prerequisite for this unit. Answers appear at the end of the unit.

This unit will examine the graphs of exponential functions (to include base e) and their inverses, logarithmic functions. All graphs should be displayed on the ZDecimal x 2 viewing window: i.e. [-9.4, 9.4] by [-6.2, 6.2], with both scales set to 1 for TI-82/83/83plus calculators. (This window will be [-12.6,12.6] by [-6.2,6.2] on the TI-85/86.)

1.   An exponential function is a function of the form $Y = f(X) = b^X$ where b is a positive real number not equal to 1.

2.   Examine the function when b (the base) is larger than 1, b > 1. GRAPH and TRACE each function below. Sketch the display carefully.

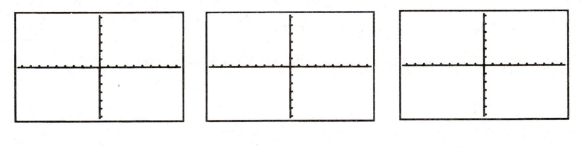

| $Y = 2^X$ | $Y = 3^X$ | $Y = 8^X$ |

✍3.   Access the TABLE (increment by 1). Scroll up and down while examining the values for Y. This can also be accomplished by carefully examining X and Y values while tracing.

a. As X increases, what happens to Y?

b. As X decreases, what happens to Y?

c. Will the value of Y ever be equal to 0? Why or why not?

d. State the domain and the range of the functions.

   Domain:_____   Range:_____

e. How are the above 3 graphs

   similar?:

   different?:

4. Examine the function when b (the base) is larger than 0 but less than 1, $0 < b < 1$. GRAPH and TRACE each function below. Sketch the display carefully.

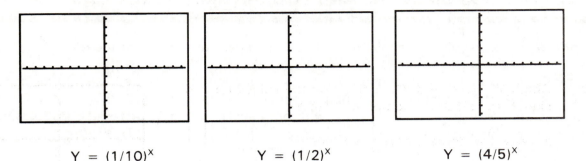

$Y = (1/10)^X$ $Y = (1/2)^X$ $Y = (4/5)^X$

✍5. Access the TABLE (ensure that it is set to increment by 1) or use TRACE. Scroll up and down while examining the values for Y.

a. As X increases, what happens to Y?

b. As X decreases, what happens to Y?

c. Will the value of Y ever be equal to 0? Why or why not?

d. State the domain and the range of the functions.

Domain:_____   Range:_____

e. How are the above 3 graphs

similar?:

different?:

✍6. The exponential functions graphed thus far are continuous. Moreover, they are either continuously increasing or continuously decreasing. Explain what the word "continuous" means in the above contexts, i.e. define continuous, continuously increasing function, and continuously decreasing function.

7. All of the graphs above have the same Y-intercept. What is it? _____

8. Why is the point with coordinates (0,1) on each of the above graphs?

144

✎9. If an exponential function is multiplied by a constant, what do you THINK will happen to the graph?

10. Graph $Y1 = 2^X$ and $Y2 = 6(2^X)$ on the calculator and sketch the display. Compare Y2 to Y1.

    a. Was the domain or range affected?_____

    b. Did the Y-intercept change?_____

    c. What is the Y-intercept of Y2?_____

11. Predict the Y-intercept for $Y = 8(2^X)$._____

12. What if a constant were added to the function $Y = 2^X$? Describe the manner in which the constant will shift the graph.

    positive constant:

    negative constant:

13. Graph the exponential functions $Y = 2^X + 3$ and $Y = 2^X - 3$ on the calculator and sketch the display. Were your predictions from #12 correct?

14. Graph the exponential function $Y = 4^X - 18$ on the calculator and sketch the display.

    a.   State the domain:_____

    b.   State the range:_____

    c.   State the Y-intercept:_____

    d.   State the X-intercept:_____

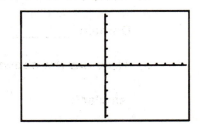

15. Consider the **equation** $4^X = 18$. Graphically solve this equation using the ROOT/ZERO feature of the CALC menu.   Circle the root on the displayed graph and record the solution: X = _____

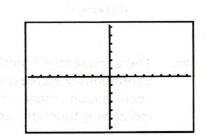

16. To solve this same equation algebraically will require the use of the Change of Base
Formula: $\log_b n = \dfrac{\log n}{\log b}$.

$$4^X = 18$$
$$X = \log_4 18$$
$$= \dfrac{\log 18}{\log 4}$$
$$\approx 2.084962501$$

Compare the approximation above with the root to the equation in #15 and the X-intercept of the graph of the equation in #14. They should all have the same approximation.

EXPONENTIAL INVERSES

The inverse of an exponential function is a logarithmic function. To find the inverse of the exponential function ($Y = b^X$) interchange the X and Y variables. This yields the equation $X = b^Y$ ($b > 0$, $b \neq 1$) which is defined as the logarithmic function $Y = \log_b X$. Until your text addresses Properties of Logarithms, you will not have an algebraic method for solving this equation for Y. Until that time, the **DrawInv** option from the DRAW menu will be used to draw the inverses of these functions.

17. Graph $Y = 2^X$. Following the instructions in the INVERSE section of the unit entitled "Functions", draw the inverse of Y1. Sketch the final display of both functions and label them as Y1 and INV Y1 on the graph.

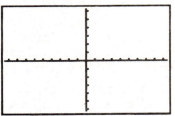

Y1: Domain:_____ Range:_____

INV Y1: Domain:_____ Range:_____

✍18. Why are the domain and range in the graph of INV Y1 the reverse of the domain and range of the Y1 graph?

Recall that the Y1 graph had a Y-intercept of 1 and no X-intercept. Although you cannot TRACE on INV Y1, what appear to be the X and Y intercepts for the **inverse of Y1**? X-intercept:_____ Y-intercept:_____

19. The defined function $Y = 2^X$ has as its inverse function $X = 2^Y$. This equation must be solved for Y in order to actually graph the inverse function as opposed to merely drawing the function. Logarithmic notation and properties allow us to do that. Solving $X = 2^Y$ using logarithmic properties yields:

$$X = 2^Y$$
$$\log X = \log 2^Y \qquad \textit{(take the logarithm of both sides)}$$
$$\log X = Y(\log 2) \qquad \textit{(Logarithmic Property: } \log_b P^n = n\log_b P)$$
$$\dfrac{\log X}{\log 2} = Y$$

146

20. Graph $Y = \dfrac{\log X}{\log 2}$ and verify the X and Y intercepts you stated in #18.

21. Use the TABLE feature to quickly complete the following tables for $Y = 2^X$ and $X = 2^Y$ and thus verify that the X and Y values of the ordered pairs are reversed in functions that are inverses of one another.

$Y = 2^X$

X	Y
1	
2	
3	
4	

$X = 2^Y$

X	Y
2	
4	
8	
16	

22. Graph $Y = 0.5^X$. Use either the DRAW menu to display the inverse of $Y = 0.5^X$ or graph the inverse of $Y = 0.5^X$ using logarithmic properties to solve for Y. Sketch the final display of both functions and label them as Y1 and INV Y1 on the graph.

Y1: Domain:_____ Range:_____

Y-intercept:_____

INV Y1: Domain:_____ Range:_____

X-intercept:_____

BASE- e EXPONENTIAL FUNCTIONS

The irrational number **e** occurs in the mathematical modeling of natural events. Its value is approximately equal to 2.71828182846, and it is often used as the base of an exponential function.

✍23. Graph $Y = 2^X$, $Y = e^X$ and $Y = 3^X$ on the screen at the right. Why does the graph of $Y = e^X$ rise faster than the graph of $Y = 2^X$ but not as fast as the graph of $Y = 3^X$?

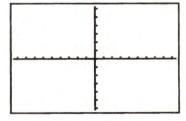

24. A common occurrence of the use of **e** in mathematical modeling is in the computation of compound interest, specifically continuous compounding. When an amount of money invested grows exponentially, the formula for computing the value of the investment is $A = Pe^{rt}$ where interest is compounded continuously.

25. If an initial investment of $2000 is placed in an account earning 4.5% interest compounded continuously, write an equation that will model the value of the investment at the end of X years.

_____

26. Graph the equation from #25 on the screen at the right. (Viewing **WINDOW HINT**: Since X represents the number of years, set the Xmax and Xmin to display non-negative values of X. To make the interpretation of the graphical display "friendlier", you may want to use the space width formulas discussed in the unit entitled "Preparing to Graph". Remember, the initial investment is $2000; this will affect the Ymax.)

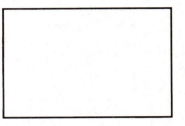

27. TRACE along the graph and determine the value of the investment (to the nearest dollar) after

1 year_____   2 years _____   5 years _____   10 years _____

✍28. Explain how to use the TABLE to determine value of the investment after 3 years 3 months.
Record this value: _____

✍29. IN YOUR OWN WORDS: Summarize what you have learned in this unit. Your summary should:
   a. state the definition of an exponential function,
   b. discuss the graphs of exponential functions when b > 0 (your discussion should address domain, range and Y-intercept)
   c. discuss the graphs of exponential functions when 0 < b < 1 (your discussion should address domain, range and Y-intercept)
   d. discuss why the case of b = 1 is excluded in the definition of an exponential function
   e. state the definition of a logarithmic function (i.e. the inverse of an exponential function)
   f. discuss the relationship between the domain, range and intercepts of an exponential function and its inverse.

Solutions:

2.

3a. As X increases, Y increases. **3b.** As X decreases, Y decreases.

3c. No: a non-zero base raised to any power is a non-zero number. **3d.** D:$\mathbb{R}$, R: $\{Y\,|\,Y>0\}$
3e. Answers may vary.

4.

5a. As X increases, Y decreases. **5b.** As X decreases, Y increases.

5c. No: a non-zero base raised to any power is a non-zero number. **5d.** D:$\mathbb{R}$, R: $\{Y\,|\,Y>0\}$
5e. Answers may vary.
6. A continuous function is one in which there are no gaps (or holes). Continuously increasing and/or decreasing functions have no turns (i.e. no relative maximums or minimums).
7. 1 **8.** When the exponent is zero the value of the exponential expression is one.
9. Answers may vary. **10a.** No **10b.** Yes **10c.** 6 **11.** 8 **12.** positive constant: shifts graph up; negative constant: shifts graph down
13.

14. D:$\mathbb{R}$, R: $\{Y\,|\,Y>-18\}$, Y-intercept: -17,
X-intercept: 2.084962501

15. ROOT: 2.084962501   **17.** Y1:see 3d.; INV Y1: domain:$\{X\,|\,X>0\}$, range:$\mathbb{R}$

18. Because the X and Y variables were interchanged. The X-intercept = 1 and there is no
Y- intercept.

22. Y1: same as 5d., Y-intercept = 1; INV Y1: domain:$\{X\,|\,X>0\}$, range: $\mathbb{R}$, X-intercept = 1

25. Y = $2000e^{.045x}$

26. WINDOW: Xmin = 0, Xmax = 94, Xscl = 0, Ymin = 0, Ymax = 5000, Yscl = 0

27. 1yr = 2092, 2yr = 2188, 5yr = 2505, 10yr = 3137

28. The TABLE will need to be incremented by .01 since 3 yrs. 3 mos. is 3.25 years. The value after 3.25 yrs. is $2315.
29. Answers may vary.

UNIT #23
PREDICT - A - GRAPH

*Unit 21 is a prerequiste for this unit. Answers appear at the end of the unit.

I.   Match the graph with the appropriate equation. Each graph in this section is graphed in the standard viewing window.

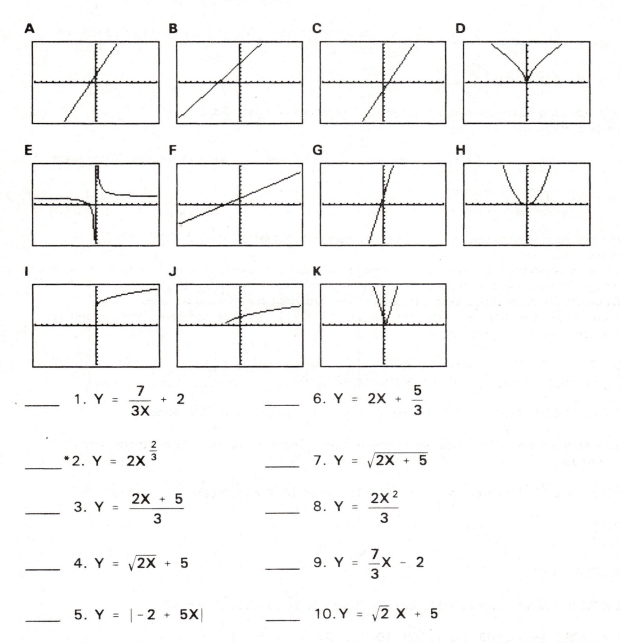

A   B   C   D

E   F   G   H

I   J   K

_____ 1. $Y = \dfrac{7}{3X} + 2$

_____ *2. $Y = 2X^{\frac{2}{3}}$

_____ 3. $Y = \dfrac{2X + 5}{3}$

_____ 4. $Y = \sqrt{2X} + 5$

_____ 5. $Y = |-2 + 5X|$

_____ 6. $Y = 2X + \dfrac{5}{3}$

_____ 7. $Y = \sqrt{2X + 5}$

_____ 8. $Y = \dfrac{2X^2}{3}$

_____ 9. $Y = \dfrac{7}{3}X - 2$

_____ 10. $Y = \sqrt{2}\, X + 5$

*NOTE: If your graph does not correspond to one of the given ones, see the solution key for assistance in entering the problem.

151

II. Match each equation to its appropriate graph. <u>Viewing</u> <u>rectangles</u> <u>may</u> <u>vary</u>. The X and Y scales are indicated if they are not equal to 1.

_____ 1. $Y = |X + 2|$

_____ 2. $Y = X^4 - 6X^2 + 9$

_____ 3. $Y = -\sqrt{2 - X}$

_____ 4. $Y = X^2 + 2$

_____ 5. $Y = |X| + 2$

_____ 6. $Y = \sqrt{X} + 2$

_____ 7. $Y = X^3$

_____ 8. $Y = \sqrt{X + 2}$

_____ 9. $Y = (X + 2)^2$

_____ 10. $Y = -X^3 - 1$

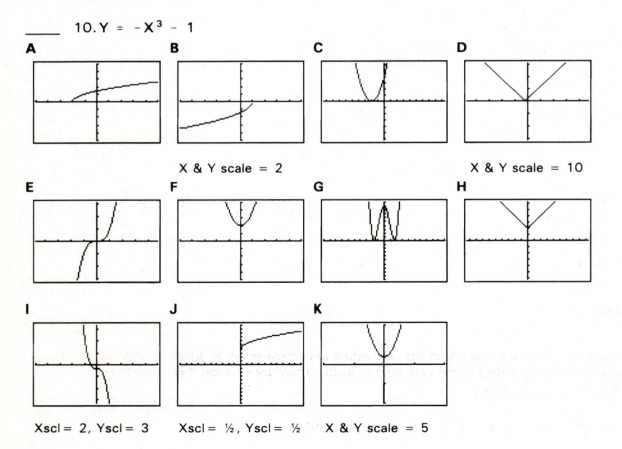

A B C D

X & Y scale = 2   X & Y scale = 10

E F G H

I J K

Xscl = 2, Yscl = 3   Xscl = ½, Yscl = ½   X & Y scale = 5

Solutions:

I. **1.** E

2. D, Because of the way several of the calculators are programmed to graph, it is best to enter rational exponents according to the definition of such. Thus, try entering this problem as

$$Y = 2(\sqrt[3]{X})^2 \quad \text{or} \quad Y = 2(\sqrt[3]{x^2})$$

3. F   **4.** I   **5.** K   **6.** A   **7.** J   **8.** H   **9.** C   **10.** B

II. **1.** D   **2.** G   **3.** B   **4.** F   **5.** H   **6.** J   **7.** E   **8.** A   **9.** C   **10.** I

*Unit 17 is prerequisites for this unit. Answers appear at the end of the unit.

Descriptive statistics is the collection, presentation and summarization of data. This unit examines paired data information and the type of plots available on the calculator for displaying this data.

1.   The STAT feature can be used as a tool for graphing ordered pairs of numbers. Unlike the graph screen, or TABLE feature, STATPLOT allows the plotting of individual ordered pairs of numbers. The graph screen and TABLE feature are dependent on functions being entered on the **Y =** screen.
        NOTE: Before proceeding, delete all entries on the **Y =** screen.

TI-85/86	ACCESS THE **y(x) =** SCREEN BY PRESSING **[GRAPH]** **[F1](y(x) =)** AND CLEAR ALL ENTRIES.

2.   The ordered pairs (2,3), (-5,7), (4,-2), (0,-4) and (5,8) can be arranged in a table of values:

X	Y
2	3
-5	7
4	-2
0	-4
5	8

Two lists will be created in the **STAT** menu to represent each list in the table of values.

TI-85	TI-85 USERS GO TO THE APPENDIX (PG.161).

TI-86	TI-86 USERS GO TO THE APPENDIX (PG.163).

3.   Begin by pressing **[STAT]** and **[1:Edit]**. Use the ► arrow key to cursor over to the sixth list (L6) to observe the content of each of the lists. Data must be cleared from the lists before beginning any problem.

4.   There are several approaches to clearing lists. For demonstration purposes, suppose that only the first and second lists (L1 and L2) are to be cleared. To clear these lists, press **[2nd]** **<QUIT>** to return to the home screen, press **[STAT]** **[4:ClrList]** **[2nd]** **<L1>** **[,]** **[2nd]** **<L2>** **[ENTER]**.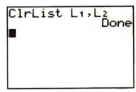

155

Another approach is to press [STAT] [1:Edit], use the ▲ key to highlight L1 and press [CLEAR]. Cursor down below L1 to see the list clear.

These approaches are useful for clearing specific lists while leaving other lists intact. If the intent is to clear **all** lists then the quickest approach is to press [2nd] [MEM] <4:ClrAllLists> [ENTER].

> **TI-82** Press [2nd] <MEM> [2:Delete] [3:List] and placing the marker in front of each list press [ENTER].

5.   The data (ordered pairs in the table of values) can be entered in the lists on the **Edit** screen in two ways. Examine the first approach by using the data listed in #2.

a. Press [STAT] [1:Edit]. Enter each X value, in order, in the L1 list.
(Press [2] [ENTER] [(-)] [5] [ENTER] [4] [ENTER] [0] [ENTER] [5] [ENTER].)

b. Use the ► arrow key to cursor to the L2 list and enter the Y values. Double check the paired data to be sure they are the same as the original table of values.

c. Press [2nd] <STATPLOT> [1:Plot 1]. Press [ENTER] to turn on the first plot. Use the ▼ arrow key to highlight the first icon after **Type** and press [ENTER].

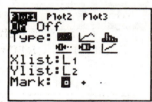

> **TI-82** Use the ▼ arrow key to highlight the first entry (L1) after **Xlist** (press [ENTER]), the ▼ and ► arrow keys to highlight the second entry (L2) after **Ylist** (press [ENTER]), and finally, the ▼ arrow key to highlight the first entry after **Mark** (press [ENTER]).   At this point, continue reading at letter d.

Use the ▼ arrow key followed by [2nd]<L1> to enter **L1** after **Xlist** if L1 is not already displayed. Then use [▼] to **Ylist** and press [2nd] <L2> to display **L2** after **Ylist**. Finally, use the ▼to highlight the first entry after **Mark**. Plotted data points can be represented by □, +, or a •. We selected □ for easy visibility.

d. Set the standard viewing window to view the ordered pairs.

e. Press [TRACE]; the screen at the right is displayed. P1:L1,L2 is displayed in the upper left corner of the screen, indicating that Plot 1 is turned on and the ordered pairs are graphed from lists L1 and L2. The ◄ and ► arrow keys allow movement from point to point, displaying the ordered pair values at the bottom of the screen. The display is called a Scatter Plot.

> **TI-82** Displayed in the upper right corner of the screen will be **P1**, indicating that Plot 1 is turned on.

f. Ordered pairs can be plotted using either the **ZOOM** menu, or by entering values on the **WINDOW** screen. If entering values on the WINDOW screen, be sure that the X values in the L1 list are between the Xmin and Xmax of the WINDOW and that the Y values in L2 list are between the Ymin and Ymax of the WINDOW.

g. Press **[ZOOM] [9:ZoomStat]** and the points are replotted with the calculator automatically resetting the viewing WINDOW to include all the data points.

6. Clear lists **L1** and **L2** using one of the methods in #4.

7. A new table of values will now be entered using a different approach.

a. The table of values is

X	Y
3	6
-2	1
4	-2

To enter the entire set of values at once, return to the home screen. Use the **{ }** to list all the X values and **STO**re them in L1.

To accomplish this, press **[2nd] < { >** **[3] [,] [(-)] [2] [,] [4] [2nd]** **< } > [STO▸] [2nd] <L1> [ENTER]**.

TI-85/86

TI-85: THE BRACES, **{ }**, ARE FOUND UNDER THE **LIST** MENU BY PRESSING **[2ND] < LIST >**. TO LOCATE **xStat** AND **yStat**, PRESS **[2ND] < VARS > [MORE] [MORE] [F3](STAT)** AND USE THE ▼ KEY TO PLACE THE MARKER AT **xStat**. PRESS **[ENTER]** TO DISPLAY THE SCREEN AT THE RIGHT.

{3, -2, 4}→xStat
                {3 -2 4}

TI-86: THE BRACES, **{ }**, ARE FOUND UNDER THE **LIST** MENU BY PRESSING **[2ND] < LIST >**. **xStat** AND **yStat** CAN BE ACCESSED FROM THE LIST MENU BY PRESSING **[F3](NAMES)**.

---

b. Store the list of Y values in L2. Your screen should now look like the one at the right.

{3, -2, 4}→L₁
                {3 -2 4}
{6, 1, -2}→L₂
                {6 1 -2}

c. Check **Plot 1** under the **STATPLOT** menu to be sure all items are selected as in #5c.

d. Press **[ZOOM] [9:ZoomStat]**. Sketch the graph displayed.

TI-85/86

TI-85: THERE IS NO FEATURE COMPARABLE TO ZoomStat. **RANGE** VALUES MUST BE SET BY HAND.

TI-86: PRESS **[GRAPH] [F3](ZOOM) [MORE] [F5](ZDATA)**.

NOTE: Remember to clear the lists in **L1** and **L2** before beginning a new problem. (See #4 for directions.)

8. Enter the table of values in **L1** and **L2** under the **STAT/EDIT** menu and display the data points using the **ZoomStat** viewing WINDOW. Sketch the graph displayed.

X	Y
1	3
2	-5
0	0
-1	-2
-3	2
5	1

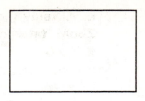

9. **Application:** On the first five days of January, the daily highs were 22°, 19°, 12°, 17° and 13°. Let days 1 through 5 be the X values and the temperatures be the Y values.

    a. Display the data points using the **ZoomStat** Window. Sketch the display.

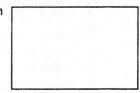

    b. Now display this same data, connecting the data points with a straight line. This is called the **XYline**. To do this, return to the **STATPLOT** menu by pressing [2nd] <STATPLOT> and [1:Plot1]. Plot1 is already turned ON, cursor down to **TYPE** and right once, to the second icon. Press [ENTER] to select the icon (which represents the **XYline**). Press [GRAPH] and sketch your screen display. This **XYline** graph clearly displays the temperature pattern over the 5 day period.

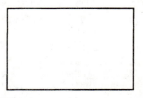

TI-86 TO DISPLAY THE XYLINE PRESS [2ND] <STAT> [F3](PLOT) [F1](PLOT1), CURSOR DOWN TO TYPE AND PRESS [F2](XYLINE). PRESS [GRAPH] TO DISPLAY THE GRAPH.

10. Using {-3, -2, -1, 0, 1, 2, 3} for the list of X values, create a table of values for the equation Y = 3X - 2. Use the **STOre** feature to accomplish this. Using braces, enter the list of X values on the home screen and store in L1. Because the X values are now in L1, evaluate 3L1 - 2 (which will create the values for the list L2) and store those evaluations in L2. Remember, colons are used to separate each command. Before pressing "enter", your screen should look like the one at the above right. Press [ENTER] to activate the command then press [STAT] [1:Edit] to view the lists L1 and L2.

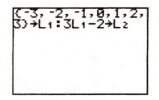

TI-85/86 THE TI-85/86 SCREEN WILL LOOK LIKE THE ONE AT THE RIGHT BEFORE PRESSING "ENTER" TO ACTIVATE THE COMMAND. REFER TO THE TI-85/86 NOTE IN #7A FOR DIRECTIONS ON LOCATING THE STAT VARIABLES.

a. Display the data points as a **Scatter Plot** graphed in the **ZoomStat** viewing WINDOW and sketch the screen display.

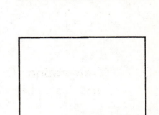

b. Now display this same data with an **XYline**. Sketch the screen display.

✍11. IN YOUR OWN WORDS: summarize what you have learned in this unit. Include the following:
a. how to clear lists in **STAT**
b. how to enter data in the lists
c. how to select the Plot display
d. explanation of the difference between Scatter Plots and XYlines.

Solutions: 7d.   8.   9a.   9b.   10a.   10b.

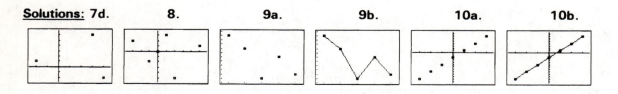

11. Answers may vary.

3. Begin by pressing **[STAT]** and **[F2](EDIT)**. Each list in the TI-85 must be named (notice the cursor is a blinking "A" indicating alpha mode). The default names for the first two lists are **xStat** and **yStat**. Press **[ENTER]** twice to accept these names for the first two lists of data. Each list must have a different name.

4. The list can be cleared by pressing **[STAT] [F2](EDIT) [ENTER] [ENTER]** to display the lists xStat and yStat followed by **[F5](CLRxy)**.

5. a. Data must be entered pairwise (data from #2 on pg. 155), pressing **[ENTER]** after each entry except the last. Displayed on the screen at the right are the first two ordered pairs of values.

    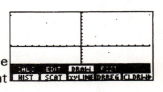

    Data can be sorted in ascending order based on either the x value or y value. Press **[F3](SORTX)** to order the pairs based on the x values.

    b. Pressing **[2nd] <M3>(DRAW)** displays the submenu that allows the drawing of scatter plots (SCAT), xyLINEs, and histograms (HIST). Press **[F2](SCAT)** to display the scatterplot at the right. It may be necessary to press **[EXIT]** ONCE to remove the top menu line or **[CLEAR]** to remove both menu lines so that the entire scatterplot is visible. These points are displayed in the standard viewing window.

    Because these points are activated from the **DRAW** menu, **TRACE** cannot be used. The TI-85 is not graphing these points. The viewing window (**GRAPH/RANGE**) must be determined by hand, being sure that these values fall between xMin and xMax and that the y values fall between yMin and yMax. There is no ZoomStat feature on the TI-85 nor are there any options such as the □ or + for the scatterplot display.

6. Press **[F5](CLDRW)** to clear the drawn pixel points before proceeding to the next problem.

☞ RETURN TO THE CORE UNIT STEP #7 (PG.157) AND COMPLETE THE UNIT .

3. Begin by pressing **[2nd] <STAT>** and **[F2] (EDIT)**. The three lists displayed must be cleared of data before beginning any problem.

4. To clear the lists, use the up arrow key to highlight **xStat** (or **yStat** or **fStat**) and press **[CLEAR]**. Cursor down into the list to see it actually clear.

5a,b. Data is entered item by item as in the TI-82 (items #5a and b from the main unit) with the only difference being that the lists are named **xStat** and **yStat** rather than **L1** and **L2**. Enter the data from page 155 (#2) now by entering each value in the x column, pressing **[ENTER]** after each entry. Cursor over to the yStat column and enter the y-values, pressing **[ENTER]** after each entry.
    Note: Data can be sorted in ascending order based on either the **xStat** values or **yStat**. Return to the home screen. Press **[2nd] <LIST>** **[F5] (OPS)** **[MORE] [MORE]** **[F5] (Sortx)**. The prompt displayed is **Sortx**. Enter a parenthesis and press **[2nd] <M3>(NAMES)**. The list to be sorted is entered first and the list that corresponds to it is entered second. Press **[F2] (xStat)** **[,]** **[F3] (yStat)** **[ )]** **[ENTER]**. The word **DONE** should appear to indicate the process is complete. Return to **STAT EDIT** screen to view list.

5c. Press **[2nd] <STAT>** **[F3] (PLOT)** **[F1] (PLOT1)** **[ENTER]** to highlight **ON** on the **PLOT1** screen. Your **STAT PLOT** is now on. Cursor down to **Type** and press **[F1] (SCAT)**; cursor down to **Xlist Name** and press **[F1] (xStat)**; cursor down to **Ylist Name** and press **[F2] (yStat)**; cursor down to **Mark** and press **[F1] (□)**. Press **[GRAPH] [F3] (ZOOM) [MORE]** and **[F5] (ZDATA)** to display the scatter plot of the data. Because the data was displayed from the **GRAPH** menu, you can use the **TRACE** feature.
    Note: The data can be displayed from the **DRAW** menu in the following manner: Return to the Home Screen. Press **[2nd] <STAT>** **[F4] (DRAW)**. Your display can be changed from this menu only by pressing **CLDRW** and returning to the defaults you set in **PLOT**.

6. Clear the xStat and yStat lists as directed in #4.

☞ RETURN TO THE CORE UNIT STEP #7 (PG.157) AND COMPLETE THE UNIT .

UNIT 25
FREQUENCY DISTRIBUTIONS

*Unit 24 is a prerequisite for this unit. Answers appear at the end of the unit.

This unit will continue to look at methods for presenting and summarizing data. The last unit examined scatter plots and XYlines as means of presenting two variable statistics. One variable data can be presented as a histogram or a box-and-whisker plot. A histogram is a vertical bar graph with no spaces between the bars. The horizontal axis (X-axis) displays the values (L1) and the vertical axis (Y-axis) displays the frequencies (L2) at which the L1 values occur. A box-and-whisker plot summarizes data using the extreme data values, the median and the upper and lower quartiles.

CONSTRUCTING HISTOGRAMS

The following steps should be followed when displaying one variable statistics as a histogram. Read through these steps: **DO NOT** implement the steps until you have been given a problem with data.

a.   Construct a frequency table of the data:

b.   Delete entries on the **Y =** screen and clear all lists L1- L6 by pressing **[2nd] <MEM>** **[4:ClrAllLists].**

> **TI-82**   Press **[2nd] <MEM> [2:Delete] [3:List]**, place the marker beside each list, press **[ENTER]** to delete.

> **TI-85**   AFTER DELETING ALL y(x) = ENTRIES, RETURN TO THE HOME SCREEN BY **[2ND] <QUIT>**. CLEAR THE **XLIST** AND **YLIST** BY PRESSING **[STAT] [F2](EDIT) [ENTER] [ENTER]**, FOLLOWED BY **[F5](CLRxy)** UNTIL ONLY x_1 AND y_1 ARE DISPLAYED.
> ON THE TI-85, THE **YLIST** WILL BE DESIGNATED AS THE FREQUENCY LIST. SKIP STEPS 3 AND 4 (THE INFORMATION REQUIRED IN STEP 4 WILL BE APPLIED WHEN YOU ARE READY TO DISPLAY THE 1-VARIABLE STATISTICS) AND PROCEED AT STEP 5.

> **TI-86**   CLEAR THE LISTS AS DIRECTED IN #4 IN THE TI-86 GUIDELINES ON PAGE 163.
> PRESS **[2ND] <STAT>   [F3] (PLOT)   [F1] (PLOT1)   [ENTER]** TO HIGHLIGHT **ON**. CURSOR DOWN TO **TYPE** AND PRESS **[F4] (HIST)**. CURSOR DOWN TO **XLIST NAME =** AND PRESS **[F1] (xSTAT)** TO ENTER **xSTAT** AS YOUR FIRST DATA LIST. CURSOR DOWN TO **FREQ =** AND PRESS **[F2] (ySTAT)** TO ENTER **ySTAT** AS YOUR FREQUENCY LIST. PROCEED TO STEP #5.

c.   Turn on **STATPLOT** and set up for a histogram. Press **[2nd]** **<STATPLOT> [1:Plot 1]** (**ON** is highlighted), cursor to **Type** and highlight the third icon (fourth icon on the TI-82), press **[ENTER]**, cursor down and ensure L1 is specified for **Xlist** and L2 for **Frequency**. Your screen display should correspond to the one at the right. (TI-82 note follows.)

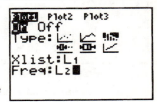

Set the calculator for one variable statistics by pressing **[STAT] [▶]** (for CALC menu) **[3:SetUp]**. L1 should be highlighted for the **Xlist** and L2 for the **Frequency**. Use the cursor to highlight the desired entry and press **[ENTER]**.

d.   Enter the data from your frequency table into the calculator's STAT list display by pressing **[STAT] [1:Edit]** and entering the values in the L1 column and their respective frequencies in the L2 column.

e.   Set the WINDOW: Be sure that the X values in the frequency table fall between the Xmin and Xmax, that the Ymin = 0 and Ymax is greater than the largest frequency value. The Xscl will determine how many values are grouped in each vertical bar. The calculator **requires** that $\frac{Xmax - Xmin}{Xscl} \leq 47$. Set the Xscl = 1 unless otherwise noted. Yscl will equal 1 since it represents frequencies. Press **[GRAPH]**. You may TRACE on the histogram; note that the number **n** will indicate how many values are in the bar that the trace cursor is located on.

TI-85/86   THESE CALCULATORS REQUIRE THAT $\frac{xMax - xMin}{xScl} \leq 63$.

f.   Translate the calculator's histogram to a hand drawn and _labeled_ histogram.

Example 1:   A 10 point quiz in an intermediate algebra class yielded the following scores: 10,9,6,7,8,8,9,8,7,7,6,7,8,8,8,7,9

Construct a histogram to represent the frequency distribution. **(The lettered steps correspond to each of the lettered steps listed previously.)**

Solution:

a.   Begin by constructing a frequency table:

X	10	9	8	7	6
frequency	1	3	6	5	2

b.   Clear all lists L1 through L6.

TI-85/86   TI-85 USERS SKIP STEPS LETTERED C AND D.
TI-86 USERS SKIP ONLY STEP LETTER D.

c.   Turn on **STATPLOT 1** and set up for a histogram display.

TI-82 users set the calculator for one variable statistics by pressing **[STAT] [▶]** (for CALC menu), **[3:SetUp]**. L1 should be highlighted for the **Xlist** and L2 for **Frequency**. Use the cursor to highlight the desired entry and press **[ENTER]**.

d.   Enter the data from your frequency table into the calculator's STAT list display by pressing **[STAT] [1:Edit]** and entering the grades in the L1 column and their respective frequencies in the L2 column.

e. Set the WINDOW values as follows: Xmin = 6, Xmax = 11, Xscl = 1, Ymin = 0, Ymax = 7, Yscl = 1. Press **[GRAPH]** to display the histogram at the right.

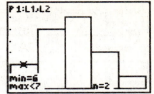

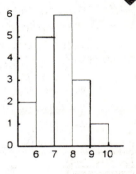

TI-85	PRESS **[GRAPH] [F2](RANGE)** TO ENTER THE WINDOW VALUES. RETURN TO THE **STAT/DRAW** MENU AND SELECT **[F1](HIST)**. PRESS **[EXIT]** TO REMOVE THE TOP LINE OF MENU DISPLAY.

TI-86	PRESS **[GRAPH] [F2] (WIND)**, ENTER THE DESIGNATED VALUES AND PRESS **[F5](GRAPH)** TO DISPLAY THE HISTOGRAM.

Drawn by hand, the histogram would look like the one at the right. Press **[TRACE]** and trace on the calculator graphed histogram to verify that the height *n* of each bar corresponds to the height of the bars in the hand drawn histogram.

To display the statistical features of these grades, press **[STAT] [▶]** (to highlight **CALC**) and select **[1:1-Var Stats]**. Displayed on the home screen is **1-Var Stats**. The calculator is now set to display one variable statistics. Enter **(L1, L2)** after the **1-Var Stat** prompt. See the screen at the right. Pressing **[ENTER]** displays the statistical data.

TI-85	TO DISPLAY THE ONE VARIABLE STATISTICS, PRESS **[F1](CALC) [ENTER] [ENTER] [F1](1-VAR)**. THE TI-85 WILL NOT COMPUTE THE MEDIAN.

TI-86	TO DISPLAY ONE VARIABLE STATS, PRESS **[2ND] <STAT> [F1] (CALC) [F1] (ONEVA)** FOLLOWED BY **[( ]** AND **[2ND] <LIST> [F3] (NAMES) [F2] (xSTAT) [,] [F3] (ySTAT)** FOLLOWED BY **[ )]**. YOUR DISPLAY WILL READ: ONEVAR (xSTAT, ySTAT). PRESS **[ENTER]**.

Notice the arrow pointing down at the bottom of the screen. Continuing to cursor down displays more statistical information. Some of the statistical features about these grades that would be of interest are specified below:

$\bar{x} \approx 7.8$ ($\bar{x}$ is the arithmetic average or mean)

median = 8 ("Med" is the number that divides ranked data into two equal groups. Cursor down to display it.)

$\sigma x \approx 1.06$ (σx = standard deviation: measures the dispersion of the data about the mean)

A frequency distribution is not appropriate when there is a wide variance among the data. The example below illustrates this using course scores. In this case, a histogram representing the number of As, Bs, etc. would be more informative. The data will only be entered in L1.

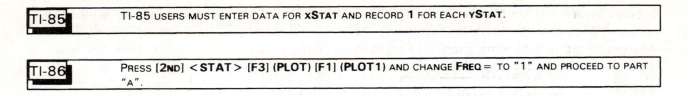
NOTE: Data can be sorted in ascending order by pressing [STAT] [2:Sort A(] [2nd] <L1> [ENTER]. When you have frequencies listed in L2, you would need to enter Sort A(L1,L2) so that the frequencies are sorted with their respective values.

If you do not use a frequency distribution then the frequency value must be set to **1** under **STATPLOT 1** and **STAT/CALC SetUp.** (Do this now.)

Example 2: The scores below are the final course scores for a college mathematics class:
83, 78, 82, 91, 84, 75, 84, 96, 80, 68, 79, 78, 85, 89, 95, 82, 88, 74, 76, 73, 74, 64, 66, 93, 69, 87
Set up a histogram to represent the grade distribution.

Solution: The local college uses a 10 point grading scale (i.e. A: 90-100, B: 80-89, etc.) and the public schools use an 8 point grading scale (i.e. A: 93-100, B: 85-92, C: 77-84, D: 69-76, F: below 69). We will construct two histograms: the distribution of letter grades for a 10 point scale and the distribution of letter grades for an 8 point scale.

10-point scale: The Xscl will determine which values (grades) are grouped in each bar. For the first histogram, set the WINDOW values as follows: Xmin = 60, Xmax = 101, Xscl = 10, Ymin = 0, Ymax = 10, Yscl = 1. Ymax may need to be increased if any one grade group has more than 10 scores in it. Press **[GRAPH].** Your displayed histogram should correspond to the pictured screen. TRACE to determine the number of As is 4, Bs is 10, Cs is 8 and Ds is 4.

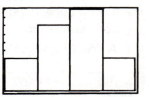

The hand drawn histogram would look like the one pictured:

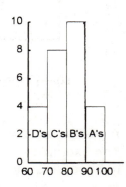

8-point scale: For your second histogram, set the WINDOW values as follows: Xmin: 60, Xmax: 101, Xscl: **8**, Ymin: 0, Ymax: 10, Yscl: 1. Ymax may need to be increased if any one grade group has more than 10 scores in it. Both your calculator generated and hand drawn sketch should correspond to those displayed.

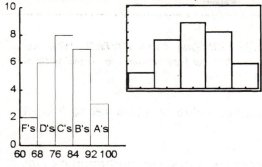

The grade distribution is as follows:
As = 3, Bs = 7, Cs = 8, Ds = 6, and Fs = 2

EXERCISE SET

Directions: For each problem complete the following:
    a. organize the data in a frequency table,
    b. enter the data in the calculator using the STAT feature,
    c. sort the data in ascending order
    d. copy the display of the histogram (Set your own WINDOW values; record the values used. Be sure to set both the Xscl and Yscl at 1.)
    e. construct a hand drawn and labeled histogram, and
    f. record the indicated information.

1. A freshmen girl's P.E. class is surveyed. The following shoe sizes were recorded:
8,7,8,8,7,7,7,6,5,6,6,6,7,9,5,7,10,9,9,8,7,7,7,6,6,8,6,7,6

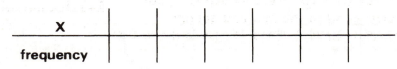

Calculator display:

WINDOW :

```
WINDOW
Xmin=
Xmax=
Xscl=
Ymin=
Ymax=
Yscl=
```

Hand Drawn (and labeled) Sketch:

$\bar{x}$ ≈ _____ (nearest whole size)

σx ≈ _____ (nearest hundredth)

median = _____

2. The daily highs for the month of February were recorded as:

68, 68, 73, 72, 73, 73, 68, 68, 73, 69, 69, 73, 73, 72, 73, 73, 68, 68, 69, 69, 69, 70, 69, 73, 70, 70, 70, 69, 71

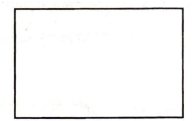

Calculator display:

WINDOW:

```
WINDOW
Xmin=
Xmax=
Xscl=
Ymin=
Ymax=
Yscl=
```

Hand Drawn (and labeled) Sketch:

What is the average temperature? _____
(to the nearest tenth of a degree)

What is the median temperature? _____

CONSTRUCTING BOX-AND-WHISKER PLOTS

TI-85 THE TI-85 DOES NOT HAVE A BOX-AND-WHISKER PLOT FEATURE.

The following steps should be followed when displaying one variable statistics as box-and-whisker plots. Read through these steps: **DO NOT** implement the steps until you have been given a problem with data.

a.   Construct a frequency table of the given data.

b.   Delete all entries on the **Y =** screen and clear all lists L1 through L6.

c.   Turn on **STATPLOT** and set up for box-and-whisker plots. Press [2nd] <**STATPLOT**> [1:Plot1] (**ON** should be highlighted), cursor down to **TYPE** and over to highlight the fifth icon, press [**ENTER**], cursor down enter L1 for **Xlist** and L2 for **Frequency**. Your screen display should correspond to the one at the right. (TI-86 note follows.)

TI-82   Under **Type**, the third icon should be highlighted for boxplot. Press [**ENTER**] to highlight L1 for **Xlist** and L2 for **Frequency**.

d. Enter the data from your frequency table into the L1 and L2 columns of the **STAT-Edit** screen.

e. Set the WINDOW: Be sure that your Xlist values (L1) fall between the Xmin and Xmax and that the Ymin and Ymax guarantee your plot will be displayed in the first quadrant of the graph.

f. Press **TRACE** and trace on the box-and-whisker plot to display the data summary.

NOTE: If data was previously graphed as a histogram, then only the **STATPLOT** screen needs to be changed to alter the way data is displayed.

Example 3: Redisplay the data in Example 1 as a box-and-whisker plot instead of a histogram.

Solution: Steps a and b will be the same as for a histogram. At step c, turn on **STATPLOT 1** and set up for a box-and-whisker plot. Steps d through e will be the same as for a histogram except that Xmin has been decreased by one so that the box plot is distinct from the X-axis. The Ymin and Ymax are not relevant to the boxplot, however, Ymin must still be less than Ymax. Press **GRAPH** to display the pictured screen.

Note: If three plots are highlighted on the **Y=** screen, then the boxplot of Plot 1 will be displayed in the upper section of the screen, Plot 2 will be displayed in the middle section and Plot 3 will be displayed in the lower section of the screen.

Press [TRACE]. The median is the first piece of information displayed. Tracing to the left yields the following two pieces of information:

Q1 = 7: The lower quartile (median of the lower half of the data) is represented by Q1.

minX = 6: This is the minimum entry in the **Xlist**.

Tracing to the right, past the median, yields the following information:

Q3 = 8.5: The upper quartile (median of the upper half of the data) is represented by Q3.

maxX = 10 This is the maximum entry in the **Xlist**.

The interquartile range can now be computed: Q3 - Q1 = 1.5.

Directions: Display the given data as a box-and-whisker plot. Record the indicated information.

3. Using the data information from EXERCISE 1 on the freshmen girl's P.E. class, sketch the box-and-whisker plot displayed. You may use the same WINDOW values as you did for the histogram.

    Median:_____

    Lower quartile:_____

    Upper quartile:_____

    Interquartile range:_____

4. Using the data information from EXERCISE 2 on the daily highs for the month of February, sketch the box-and-whisker plot displayed. You may use the same WINDOW values as you did for the histogram.

    Median:_____

    Lower quartile:_____

    Upper quartile:_____

    Interquartile range:_____

5. Using data information from Example 2 on the test scores, sketch the box-and-whisker plot displayed. You may use the same WINDOW values as you did for the histogram.

    Median:_____

    Lower quartile:_____

    Upper quartile:_____

    Interquartile range:_____

Solutions:

Exercise 1.

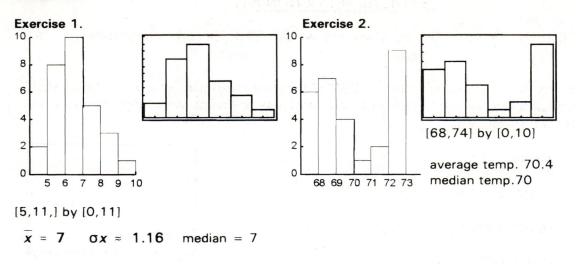

[5,11,] by [0,11]

$$\overline{x} = 7 \quad \sigma x \approx 1.16 \quad median = 7$$

Exercise 2.

[68,74] by [0,10]

average temp. 70.4
median temp. 70

Exercise 3. median = 7, lower quartile = 6, upper quartile = 8, interquartile range = 2

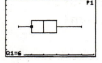

Exercise 4. median = 70,
lower quartile = 69, upper quartile = 73,
interquartile range = 4

Exercise 5. median = 81,
lower quartile = 74, upper quartile = 87,
interquartile range = 13

173

UNIT 26
LINE OF BEST FIT

*Unit 24 is a prerequisite for this unit. Answers appear at the end of the unit.

This unit determines the relationship between two variable quantities (data points) in different types of problems. If the data points lie in a straight line then the equation of the line passing through these points can be used to predict unknown data points. The line of best fit, defined by a linear regression equation, is a line that comes "close" to passing through all the data points. This unit explores only LINEAR REGRESSIONS. (In situations where the relationship is not linear, students should examine one of the following types of regression equations: quadratic, cubic, quartic, exponential or power.)

STEP-BY-STEP PROCEDURE FOR REGRESSION EQUATIONS

TI-85   TI-85 USERS GO TO THE APPENDIX (PG.181).

a.   Clear all data from lists L1 - L6 and delete all entries from the **Y**= screen.

b.   Enter paired data in lists L1 and L2.

c.   Press [**STAT**] [▶] (to highlight **CALC**) [**2:2-Var Stats**] [**ENTER**]. On the home screen "2-Var-Stats" will be displayed. Press [**ENTER**] to display statistics.

TI-82   **Press [STAT] [▶]** (to highlight CALC) [**3:Setup**]. Under 2-Var Stats highlight L1 for the Xlist and L2 for the Ylist.

TI-86   PRESS [2ND] < STAT > [F1] (CALC) [F2] (TwoVa) [( ] FOLLOWED BY [2ND] < LIST > [F3] (NAMES) [F2] (xStat) [,] [F3] (yStat) AND [ )]. PRESS [ENTER] TO ACTIVATE.

d.   Turn **ON STATPLOT 1**, highlight the first icon under **TYPE** (scatter plot), L1 for Xlist and L2 for Ylist. There are three choices for the type of Mark; select " □ " for visual clarity.

e.   Press [**ZOOM**] [**9:ZoomStat**] to view the scatter plot. If you have a mechanical pencil, use a piece of the lead - or a similarly short piece of spaghetti, to approximate the "line of best fit". The next steps will instruct the calculator to compute the equation of this "line of best fit" and graph it.

TI-86   PRESS [**GRAPH**] [**F3**] (ZOOM) [**MORE**] [**F5**] (ZDATA).

f.   Press [**STAT**] [▶] [**4:LinReg(ax + b)**] [**ENTER**].

TI-82   **Press [STAT] [▶] [5:LinReg(ax + b)] [ENTER].**

On the home screen the equation y = ax + b, along with the corresponding variable values, will be displayed. Directions for copying this equation to the **Y**= screen are in the next step.

g.   Press [VARS] [5:Statistics] [▶] [▶] (to highlight EQ for equation) [1: RegEQ].

h.   Press [GRAPH] to display the scatter plot and the "line of best fit". The viewing WINDOW was automatically set to ZoomStat in step letter e. As long as the **STATPLOT** is turned **ON** you will be able to TRACE on the scattered data points and on the line using the ▲ or ▼ keys to move from data point to line. You will not be able to TRACE beyond the ZoomStat viewing window unless you edit the WINDOW values. For this reason, we will examine information relating to the line by using the TABLE feature.

Example: The following table gives the winning time (in seconds) for the Men's 400-Meter Freestyle swimming event in the Olympics from 1904 to 2000. The Olympics were not held during the years 1916, 1940, and 1944 due to the two World Wars. Find a linear regression equation to predict the winning times for each of those years.

YEAR	1904	1908	1912	1920	1924	1928	1932	1936	1948	1952
TIME	376.2	336.8	324.4	326.8	304.2	301.6	288.4	284.5	281	270.7

1956	1960	1964	1968	1972	1976	1980	1984	1988	1992	1996	2000
267.3	258.3	252.2	249	240.3	231.9	231.3	231.2	227	225	228	213.6

(Source: *The New York Times 1998 Almanac*)

Solution: Lettered steps correspond to the numbers listed on the previous page.

a.   Clear data from lists.

b.   Enter years in L1 list and winning time in the L2 list.

c.   Go to **STAT/CALC**, press [2:2-Var Stats] and enter (L1,L2) after 2-Var Stat on the home screen and press [ENTER].

d. Turn on **STATPlot 1** (be sure all others are OFF). Your STATPLOT screen should look like the one displayed.

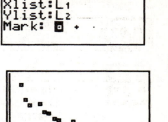

TI-82 The TI-82 display will have L1 highlighted after **Xlist** and L2 highlighted after **Ylist**.

e. Press **[ZOOM]** **[9:ZoomStat]** to view the scatter plot. Your display should correspond to the one at the right. Place your pencil lead on the screen to approximate the line of best fit. Draw in this line on the screen at the right.

TI-86 PRESS **[GRAPH]** **[F3]** (ZOOM) **[MORE]** **[F5]** (ZDATA).

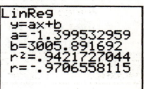

f. Determine the linear regression equation (the equation for your line of best fit) by pressing **[STAT]** **[▶]** **[4:LinReg(ax + b)]** **[ENTER]** at the home screen. The following information should be displayed on your home screen.

The correlation coefficient on the displayed screen is $r \approx -.97$. This is the measure of tendency for our variables X and Y to be related in a linear way. The fact that **r** is negative indicates our data points lie near a "falling" line. The value of r is always between -1 and 1, with values close to zero indicating that there is no tendency to a linear relationship. (Some calculator versions may not display the r^2 value.)

Note: The correlation coefficient is not automatically displayed on the regression equation screen because the calculator default mode is set to **Diagnostic Off**. One way to determine the **r** value is to press **[VARS]** **[5:Statistics]** **[▶]** **[▶]** to highlight **EQ**, and **[7:r]** **[ENTER]**. Or, if the r-value is always needed then the calculator should be reset to **Diagnostic On**: press **[2nd]** **<CATALOG>** **[▼]** until the pointer marks **Diagnostic On**, **[ENTER]** to copy the command to the home screen and **[ENTER]** again to activate the command.

TI-86 PRESS **[2ND]** **<STAT>** **[F1]** (CALC) **[F3]**(LINR) **[ENTER]**.

g. Go to the **Y =** screen. To copy this equation to the **Y =** screen, press **[VARS]** **[5:Statistics]** **[▶]** **[▶]** **[1:RegEq]**.

TI-86 PRESS **[GRAPH]** **[F1]** (Y(X)). TO COPY THE LINEAR REGRESSION EQUATION TO THE Y(X) = SCREEN, PRESS **[2ND]** **<CATLG-VARS>** **[MORE]** **[MORE]** **[F4]** (STAT), CURSOR DOWN TO REGEQ AND PRESS **[ENTER]**.

h. Press **[GRAPH]** to view the scatter plot and the line of best fit. See screen display at the right.

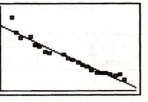

TI-86 PRESS **[2ND]** **<GRAPH>** **[M5]** (GRAPH) TO VIEW THE SCATTERPLOT AND THE LINE OF BEST FIT. NOTE: USE THE TABLE FEATRUE TO PREDICT VALUES AS INDICATED FOR THE TI-82/83.

To predict the winning times in the missing years, set the table to start at 1916 (increment by 4 - Olympics occur every 4 years). Press **[2nd]** **< TABLE >**, scroll to view the predicted winning times for the missing Olympic years. The winning time in 1916 would have been 324.39, in 1940: 290.80 and in 1944: 285.20. This regression line comes "close" to passing through the set of points. Therefore, your predictions are approximations at best.

TI-85	TI-85 users are reminded to use **evalF** - press **[2nd]** **<CALC>** - to make predictions.

EXERCISE SET

Directions: For each problem, follow the steps on the first page of this unit. Sketch the graph screen that displays the scatter plot and the calculator's line of best fit. Use the TABLE to answer the question(s) posed.

1. The chart below gives the year and winning time (in seconds) for the Men's 1000-Meter Speed Skating event. Predict the winning time for the 2006 Winter Olympics.

YEAR	1976	1980	1984	1988	1992	1994	1998
TIME	79.32	75.18	75.8	73.03	74.85	72.43	70.64

(Source: *The New York Times 1998 Almanac)*)

Graph display:

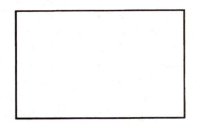

Winning time prediction for the 2006 Winter Olympics:_____

What is the value (to the nearest hundredth) of the correlation coefficient?_____

2. A survey was taken of notably tall buildings on the east coast, mid-west, and west coast. The survey compares height, in feet from sidewalk to roof, to number of stories (beginning at street level). If the Empire State building has 102 stories, use the given information to predict the height of the building.
    (Source: The World Almanac and Book of Facts 1995, actual height is 1250 ft.)

BUILDING	STORIES	HEIGHT (FT.)
Baltimore U.S. Fidelity and Guaranty Co.	40	529
Maryland National Bank (Baltimore, MD)	34	509
Sears Tower (Chicago, IL)	110	1454
John Hancock Building (Chicago, IL)	100	1127
Transamerica Pyramid (San Francisco, CA)	48	853
Bank of America (San Francisco, CA)	52	778

Graph display:

Predicted height of the Empire State Building:_____

What is the value (to the nearest hundredth) of the correlation coefficient?_____

3. The statistics in the table below represent dollars per 100 pounds for cattle. Predict the price per 100 lb. for the years 2004 - 2006.

YEAR	1940	1950	1960	1970	1975	1979	1980	1984
PRICE	7.56	23.30	20.40	27.10	32.20	66.10	62.40	57.30

1985	1986	1987	1988	1989	1990	1991	1992	1993
53.70	52.60	61.10	66.60	69.50	74.60	72.70	71.30	72.60

1994	1995	1996
66.70	61.80	58.70

Source: The World Almanac and Book of Facts 1998)

Graph display:

Predicted price per 100 lb.:
2004:_____
2005:_____
2006:_____

What is the value (to the nearest hundredth) of the correlation coefficient?_____

Solutions:

1. Winning time: 68.588, r≈-.89

2. height: 1281.1, r≈.95

3. r≈.91

2004: 81.11

2005: 82.30

2006: 83.50

PROCEDURE FOR REGRESSION EQUATION

a.   Press **[GRAPH] [F2](RANGE)** and set range values to correspond to the data to be
        entered into xlist and ylist.

b.   Clear xlist and ylist on the **STAT/EDIT** screen.

c.   Enter data into xlist and ylist.

d.   Press **[2nd] <M3>(DRAW) [F2](SCAT)** to display the scatter plot.

e.   Press **[STAT] [F1](CALC) [ENTER] [ENTER] [F2](LINR)** to display the linear
        regression equation information and the correlation coefficient. To copy this
        information to the y1 = prompt, press **[GRAPH] [F1](y(x)=) [2nd] <VARS>
        [MORE] [MORE] [F3](STAT).** Place the pointer at **RegEq** and press **[ENTER].** Press
        [2nd] <M5>(GRAPH) to graph the regression equation. (Note: Press **[CLEAR]** to
        remove all menu lines.)
        Press **[STAT] [F3](DRAW) [F2](SCAT)** to redisplay the scatter plot.

f.   Because the regression equation is GRAPHED at y1 = , it is possible to **TRACE** on the
        line.

g.   To make predictions based on the regression equation, you will need to use the **FCST**
        (forecast) option found under **[STAT] [F4].** It returns a forecasted value for x or y
        based on the current regression equation. You press **[F5]** to solve.

☞   RETURN TO THE FIRST EXAMPLE IN THE CORE UNIT (PG.175).

TROUBLE SHOOTING

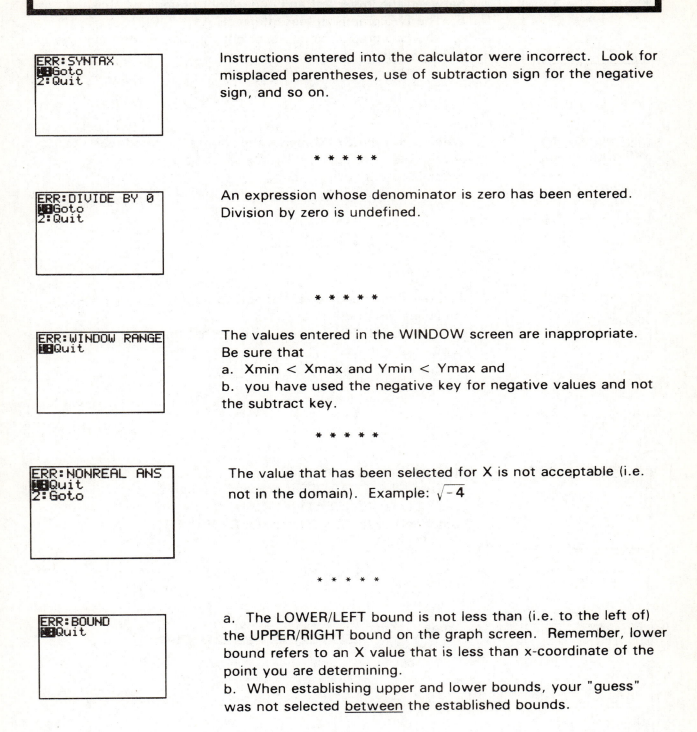

ERR:SYNTAX
1:Goto
2:Quit

Instructions entered into the calculator were incorrect. Look for misplaced parentheses, use of subtraction sign for the negative sign, and so on.

* * * * *

ERR:DIVIDE BY 0
1:Goto
2:Quit

An expression whose denominator is zero has been entered. Division by zero is undefined.

* * * * *

ERR:WINDOW RANGE
1:Quit

The values entered in the WINDOW screen are inappropriate.
Be sure that
a. Xmin < Xmax and Ymin < Ymax and
b. you have used the negative key for negative values and not the subtract key.

* * * * *

ERR:NONREAL ANS
1:Quit
2:Goto

The value that has been selected for X is not acceptable (i.e. not in the domain). Example: $\sqrt{-4}$

* * * * *

ERR:BOUND
1:Quit

a. The LOWER/LEFT bound is not less than (i.e. to the left of) the UPPER/RIGHT bound on the graph screen. Remember, lower bound refers to an X value that is less than x-coordinate of the point you are determining.
b. When establishing upper and lower bounds, your "guess" was not selected <u>between</u> the established bounds.

* * * * *

* * * * *

```
ERR:SIGN CHNG
1:Quit
```

a. No real root - graph does not intersect the X-axis between the lower/left and upper/right bounds established.
b. The two graphs do not intersect.
c. The two graphs do intersect but the intersection is not visible on the display screen.

* * * * *

```
ERR:BAD GUESS
1:Quit
```

When using the CALC menu, the "guess" entered was not acceptable. Try again, this time entering a guess as close as possible to the desired point.

* * * * *

```
ERR:DIM MISMATCH
1:Goto
2:Quit
```

a. Check dimension rules for addition, subtraction and multiplication of matrices.
b. If using the STAT menu, be sure that the lists have the same number of entries.

* * * * *

```
ERR:STAT PLOT
1:Quit
```

a. Data entered incorrectly
b. Turn STATPLOT OFF if graphing functions.
c. STAT SetUp ([STAT] [▶] (CALC) [3:SetUp]) not compatible with data lists and/or STATPLOTS.

* * * * *

```
ERR:STAT
1:Quit
```

$$\frac{Xmax - Xmin}{Xscl}$$ must be <u>less</u> <u>than</u> <u>or</u> <u>equal</u> <u>to</u> 47

* * * * *

INDEX